# Accident, Touch and Privacy:
## Climate Change and the Unguided Evolution of Humanity

Niccolo Caldararo

# Accident, Touch and Privacy:
## Climate Change and the Unguided Evolution of Humanity

Niccolo Caldararo

Academica Press
Washington~London

Library of Congress Cataloging-in-Publication Data
Names: Caldararo, Niccolo (author)
Title: Accident, touch, and privacy : climate change and the unguided evolution of humanity | Caldararo, Niccolo.
Description: Washington : Academica Press, 2025. | Includes references.
Identifiers: LCCN 2024944058 | ISBN 9781680534160 (hardcover) | 9781680534177 (e-book)

I dedicate this book to Luna
who has given us so much joy and happiness for so many years.

# Preface and Introductory Note

In his investigation of the nature of madness and civilization Foucault (1961), expresses the difficulty most historians have in addressing the points of confusion that lead a society into acts of self-destruction and chaos. He states that one problem is, "A realm, no doubt, where what is in question ·is the limits rather than the identity of a culture." The decision to continue as in the past as if no changes have occurred is a kind of madness, but the identity of normal people is associated with ideas of action defined by the culture and performed as a means of demonstrating identity. It is also an avenue for demonstrating reality both for individuals and groups, as Goffman (1959) described. Since today we have a growing world culture that is an amalgam of both European colonial masters' culture history, the defeated servants of that history and some who resisted (e,g, Japan), the nature of limits of action are embedded in the values that led to the conquest of the world. I have defined this process in detail elsewhere (Caldararo, 2022), so I will only make references as we go along to this process where necessary. However, this process remains a central motivating factor in the resistance to modifications of world economic institutions that might deflect the worst consequences of climate change for, at least, the generations to come in the next 100 years.

This is a brief outline of the chapters of this book, with a summary of the goals of the book. The title of the book includes both climate change and the words "accidental" and "evolution" for humanity's history. This is because while most cultures define their origin and purpose for existence, the evolution of humanity as a species has been subject to the vagaries of natural selection. Debate on what natural selection is or how it operates usual arrive at an agency of random survival amid a noise of struggle [see: Darwin, 1859; Spencer (1864), Wallace (1855), Dunn & Dobzhansky (1977), Gould (1977) and Dawkins (1976), etc.]. To address climate change and its threat to humanity and life on Earth as we have come to enjoy it, requires some focus on human history and culture. We have to refer to past catastrophic climate change. Some science fiction writers have attempted a description and a few journalists (e.g., Aldern, 2024) have tried to envision such a process. Many climate scientists caution that predictions will likely be flawed as past events had different contexts and factors (volcanism; asteroids; changes in solar radiation, etc.). Current evidence, for example, sets the time frame for the planet Venus' catastrophic climate change at around a billion years (Weller, et al., 2023). How fast will change happen, what will be the cumulative effects, at different times, all at once? A recent NASA compilation of data finds a climate change pattern 250 times faster than in the paleoclimate evidence (todayhttps://science.nasa.gov/climate-change/evidence/). Most important is how we have come to treat each other and in this the title reflects a social history of othering, castes, class, sex and gender, outcasts.

To cover such a social history, one must include a culture history. To understand where we are it is necessary to understand how colonialism occurred and how it shaped our social and economic conditions today. This problem I have addressed in my book (Caldararo 2022) which traces the origins of institutions to Greece, Egypt, Mesopotamia, India, and China. To confront disaster, either natural or human-created, the structure and condition of the society affect must be known. As Toynbee (1946) argued, the challenge is met or overwhelms a society, based on the internal organization and relations. In regard to colonialism, from 1492 on, the failure of non-western people to organize effective responses to European invasion is remarkable. While Europe weathered external threats after the collapse of the Western Roman Empire, it was saved by the failure of the Islamic invasion in the 8th century, death of the Great Khan in the 13th century, survived the Bubonic Plague in the 14th, and nearly fell to the Turks in the 15th and 16th. The rise of European supremacy was met with few effective defenses. Japan escaped colonization by closing its borders and was aided by the Wars of the Reformation that distracted Europe long enough for it to prepare itself (Caldararo, 2016). Ethiopia, a basically Christian state at the time of the contact with the Portuguese, was able to form an alliance that held off its destruction. Turkey was battered as the Ottoman Empire for 5 centuries before a remarkable defense was constructed in the shell of its corpse by a unique leader (Ataturk) at just the last moment.

Otherwise, the world's non-western peoples were conquered quickly as they succumbed to surprise and culture shock, disease, superior weapons, and most of all, their inability to organize a sufficient response. This is both a problem of ideology and of indigenous institutions capable of acting as effective organizers to overcome disunity and leadership. Today we can see that the ideology (religious and economic) and institutional structures (both national and international) are not based on an idea of sustainability and cooperation but on exploitation, opportunity and competition. The culture history of the world, based largely on western colonial triumph and its principles are the central blocks to successful adaptations.

Ideas of the future are still frozen in the Bronze Age concepts of the destruction of the world as a project of deities or the 19th century hubris of progressivism that posits a continuous creation of new technologies to provide future wealth, food and energy. See, for example, the essays by Utterstrom, Wilkinson and Cosby in (Worster, 1988). This hobbles any constructive actions that might be of the magnitude to make a difference in time.

The changes necessary are fundamental to human behavior, not only in consumption, but in the creation of families and the expense of children. A substantial change in behavior is needed. It also defines the main problems starting with how do we engage each others' bodies in a legitimate and yet sensual manner? Are there health issues involved, issues of social solidarity? Is contact necessary for health? Are there conditions that reflect a lack of contact, and though considered in some cases and cultures pathological, are they and is depression or happiness related? What are the limits of society's control of peoples' bodies? Of man's control of women's? Can we find cross-cultural correlates? The boundaries of friendliness versus those of personal space are a central element in social communication in most all societies as Hall (1969) demonstrated. The Covid lockdown has tested these boundaries as contact with other humans is an important factor in social identification as well as individual health. This leads into the chapter on touching.

Most books on climate change describe what is happening to the environment due to assessments of human uses of fossil fuels. This book examines in detail aspects of human culture history that make it unlikely if not impossible for humans to adapt to climate change in the present vector, or to avoid it or stop it. This perspective is not unique or new, Tainter's (1988) view of human social collapse surveyed the archaeological and ethnohistorical record showing little success in dealing with such crises. This view was reinforced by research conducted by Smil (2008) and Patten (2008), but more important in that they project abilities of states to respond to crises and give them failing marks. This largely reiterates findings 50 years ago on the problem of growth and the quality of life (Wagar, 1970). Rees (2013) addresses the idea of human exhaustion of the productive capacity of the Earth and the concept of "overshoot" and indicates that the future will likely pattern as the past. It is likely we will come to know all the problems we have created and manufacture solutions but not implement them as they would require too great changes of behavior. The general aggressive response to the Covid-19 pandemic, demonstrate a lack of compliance and cooperativeness and also real anger toward those who come up with solutions and those responsible for health care.

The problems of human sexuality, of systems of prestige and dominance that rule human life are embedded so deeply in every aspect of motivations and decisions that human change is regulated to crisis and trauma. Assault on institutions is an element of human failure to adapt and is driven by irrational behavior, though with elements of messianic movements and not revitalization attempts as defined by Wallace (1956), today's populist movements are characterized by mass movements driven by media at modern institutions of government and social organization (e.g., education, medicine, etc.). Their actions and rhetoric appear more nativistic and revivalistic (de Waal Malefijt, 1968).

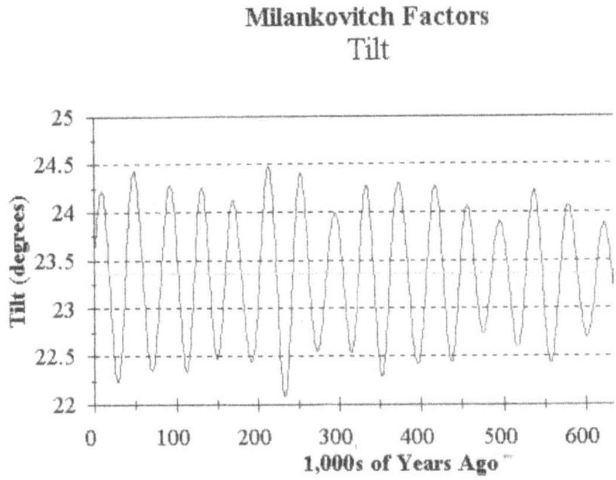


r

**Chart 1: Climate change in the past**

The book outlines facets of human behavior that could be modified under certain conditions that might provide a foundation for human survival at least into the 21st century, it also describes why this is of limited possibility. Probabilities of reward systems existing today and with long evolutionary histories produce a rigid road map for the future. How we conceive of ourselves and the fantasies and illusions surrounding our image of how we believe we are conscious of ourselves and masters of the world are mechanisms of self-defeat that underlie every possible exit.

A program of awareness needs to be constructed to guide a human realization of the futility of present social ideals that are constant hallucinations driving our destruction of the planet (Caldararo, 2023). We often hear that the idea of climate change is something natural and unavoidable, but many ecologists, for example Emmel (1977) have recognized the many ways humans create Earth disasters and facilitate natural process of destruction, as in both forest fires and flooding. We even see theories constructed that argue that some of these human created disasters are "natural." However, humans are not unique in their adaptations, neither as mammals or Primates (Caldararo, 2012). Our species has developed a ritualized system of conflict as patterned as that of other animals' intra-species conflict (Caldararo, 2004; Caldararo, 2017).

## References

Aldern, Clayton Page (2024) *The Weight of Nature: How a chanimg Climate Chanes Our Brains,* New York Dutton.

Caldararo, Niccolo (2004) "War, Mead and the nature of criticism in Anthropology," *Anthropological Quarterly*, Spring, v. 77, n, 2, 311-322.

Caldararo, Niccolo (2012) Psychic Unity of Mankind, Saarbrucken, Scholars' Press.

Caldararo, Niccolo (2022) *Origins of White Supremacy and Colonialism: Impasct of History's Interpretation on the Present*, Warsaw, Creator Publishing House.

Caldararo, Niccolo (2020) "Social complexity, pathogen adaptation and Covid-19: history of disease avoidance, social spacing and work/home matrix," *Urbanities,* v. 10, Supplement 4, 14-18.

Caldararo, Niccolo (2017) *Big Brains and the Human Superorganism,* Lanham, Rowman & Littlefield.

Crosby, Alfred W. (1988) "Ecological imperialism: the overseas migration of western Europeans as a biological phenomenon," In Donald Worster, ed. *The Ends of the Earth*, Cambridge, Cambridge University Press, 103-117.

Darwin, Charles (1859) *Origin of Species*, New York, Cassell.

Dawkins, Richard (1989) *The Selfish Gene,* Oxford, Oxford University Press, (originally published in 1976).

Dunn, L.C. and Th. Dobzhansky, (1946) *Heredity, Race and Society*, New York, Menton Books.

Emmel, Thomas C. (1977) "Introduction," in *Global Perspectives on Ecology*, (ed.) Thomas C. Emmel, Palo Alto, Mayfield Publishing Co.

Foucault, Michel (1961) *Madness and Civilization*, New York, Vintage Books.

Goffman, Erving (1959) *The Presentation of Self in Everyday Life*, New York, Anchor Books.

Gould, Steven Jay (1977) *Ontogeny and Phylogeny,* Cambridge MA, Belknap Press.

Hall, Edward T. (1969) *The Hidden Dimension,* Garden City, Anchor Books.

Patten, Chriss (2008) *What Next? Surviving the Twenty-First Century*, London, Allen Lane.

Smil, Vaclav (2008) *Global catastrophes and Trends: The Next Fifty Years*, Cambridge, MA, MIT Press.

Spencer, Herbert (1864) *Principles of Biology*, London, Williams & Northgate.

Toynbee, Arnold J. (1946) *A Study of History,* Oxford, Oxford University Press.

Utterstrom, Gustaf, (1988) "Climatic fluctuations and population problems in early modern history," in Donald Worster, ed. *The Ends of the Earth*, Cambridge, Cambridge University Press, 39-79.

Wagar, J. Alan (1970) "Growth versus the quality of life," *Science,* v. 168, 5 June, 1179-1184.

Wallace, Alfred Russell (1855) "On the law regulating the introduction of new species," *Annals and Magazine of Natural History*, 2nd series, v. 16, 196.

Weller, Matthew B., Alexander J. Evans, Daniel E. Ibarra and Alexandria V. Johnson, (2023) "Venus' atmospheric nitrogen explained by ancient plate tectonics," *Nature Astronomy* v. 7, 1436-1444.

Wilkerson, Richard G. (1988) "The English Industrial Revolution," in Donald Worster, ed. in Donald Worster, ed. *The Ends of the Earth,* Cambridge, Cambridge University Press, Cambridge, Cambridge University Press, 80-102.

# Table of Contents

These groups of young people in the 1960s and 1970s developed an approach to gender they called "gender fuck." They wanted to be free to display their sexual attitudes in any way possible, at any time and to change at will their appearance, gender and attitude. Such behavior caused a considerable problem in translating appearance (dresses, makeup and beards with long hair on men, bald heads and pants/skirts etc on women. The feeling of daily oppression by the patrifocal gaze which produced self-censorship, was held to be damaging to health and creativity. They felt the absurdity of ownership of a body part (e.g. penis) should give one person the right to enter another's body and to have considerably more social rights than one without such a body part. This leads to chapter 4 and an examination of the penis in various species, mainly birds.


This is a poorly understood fact and one that is placed in evolutionary and social context. While most male birds lack a penis, or one might say, most all genders of birds, the condition varies with reptiles where we find in some species males have two. The lack of a penis seems associated with less gender violence and more long-term relationships as well as high social engagement and investment in offspring.


How do we fit marriages of various kinds, e.g., consenting consorts, bisexual open marriages or closed ones, marriages of cis and trans men and women, genderless or new gender marriages or arrangements into social institutions? We examine past examples and explore future possible consequences, like fertility and sustainable human populations. Marriage effects fertility so it is significant in the allocation of resources.


Considerable debate on the causes of climate change hinges on a number of political and religious motives and assumptions today. Past examples of the awareness of climate change and the failure to act can provide some background to these debates. Jennings (1974) gives a comprehensive analysis of the archaeological evidence for climate change and human responses. Bryson (1973)

**Figure 1 Artificial snow keeps Swiss slopes running, humans adapt to challenge but do not change habits**

**"Climate change is personal… it is about what kind of hell your child will live in..."**

# Introduction

This is a book about being human, performing self and getting along in society. These seem simple ideas, yet they describe singly a complexity of being that has eluded social scientists and philosophers for over 2,600 years. I am not claiming this is a comprehensive assessment or history. It is a partial view from a unique perspective in the 21st century. One of the greatest threats to human existence and a sustainable world is human activity and population. The demand for resources to produce and raise children is an enormous drain on the world's resources and while Malthusian ideas of economic collapse and disaster have receded from debate (Hutchinson, 1967), human fertility is at the center of climate change. And this centers on the control of women, of their bodies and their consciousness, for if women are free to choose to produce children or to develop themselves is a key component. Luckily, it appears that human longevity and life span are limited by our biology given Olshansky, et al. (1993; 2024) studies, one 30 years after the initial findings. This will not prevent, I am sure, tech billionaires from trying to save themselves by virtual or robotic means (Sample, 2022).

There are a number of components to fertility, one is the need for a society to reproduce itself. The survival of young in a hunter-gatherer Paleolithic society was a major task of adults, but survival of young in Neolithic society seems to have been more difficult (Bogin, 2001). It is generally not a difficult task for a woman to be impregnated, and it is a mark of adulthood in many cultures and a passage in status, but holds significant liabilities and limitations (Hrdy, 1999). While most cultures require men who engage in sexual intercourse to take some responsibility for raising children, this often is shared in kinship societies. Where kinship breaks down, females are in more tenuous situations and this characterizes complex societies today. The desire for children and the commitment to raise them are often separated, as is the desire of a female to have children a complex decision, often one forced on her by circumstances. Thus no progress to dealing with climate change can be addressed without a comprehensive analysis and solution to the pregnancy issue, especially where men see getting a woman pregnant as mainly an expression of their manhood, power and status, (demonstrated as over a female) or means to an economic end.

How humans interact, develop bonds, perform genders and engage in sexual life is the focus that drives population growth. In fact, the whole question of how gender is determined not only in humans, but in any species is a fascinating subject. Hrdy (1981) questioned some accepted ideas of dominance and gender that focused on size (in sexual dimorphism), large size meaning dominant and this was often extended to gender traits. Yet when we look at the basic nature of gender, of ova and sperm, ova are millions of times larger (Johnson, et al., 2021). What about genitals? In some insects females have a penis, yet the male is termed by the ejaculate originating in him and being retained and developed in the female (Brennan, 2016). But maleness and femaleness are expressed in animal societies as different behaviors associated with dominance, violence, aggression and power. The power to control nature is one aspect. This is true in the ability to change the physiology or body shape of another, as in impregnating someone. In some animals that produce spermatophore, the males never meet females, but deposit the spermatophore (usually stalks with sperm in spheres at the end) on the ground and the females choose which to insert in their genital openings. The competition among males is then to replace the sermatophores of other males with their own as in pseudoscorpions and millipeds (Weygoldt, 1969 and Schaller, 1971).

Our gender patterns have come to shape the pattern of our exploitation of the environment. There is evidence that the loss of or reduction of reproductive organs in some insects is associated with the development of social complexity (Zhuang, et al., 2023). We celebrate the great hunter, the triumphant sports master, the takeover victor of the financial world. And while rewards of nurturing are recognized, they are defined as associated with the submissive gender in most of today's human cultures, as only patrifocal (male centered) societies have survived from the Bronze

Age when there appear to have been more matrifocal (female positive) societies. What is disturbing, but not unexpected from an evolutionary point of view, is the fact that women's bodies have evolved (like most female bodies, perhaps with some exceptions like the hyena, see Hrdy, 1981) to be attractive to the male. Some new perspectives on this have taken it to 250 million years ago (Bohannon, 2023).

In most human cultures females are trained and expected to increase this attraction by gesture, make-up and dress. Thus even if a female does not desire male attention, she is sanctioned if she does not act so. The constant bombardment of male attention (male gaze included) has often been described in detail by feminists, but the general outcome is not just male dominance, but a reinforcement of an attitude of exploitation that is at the root of habitat destruction, climate change and species extinction. We often hear discussions of free will, but in this regime of social and biological constraints one is hard pressed to suggest women have at least a modicum of it. While studies of mate choice in animals in general find that females in many species have considerable selective choice (Trivers, 1985), most human societies in modern times, provide women with few avenues for power let alone mate choice. Arguably this has changed somewhat in the past 100 years, we see fewer instances of men owning women or children, but we do see many cases of men killing women who reject them or attempt to escape their control. The ownership of children is also still difficult to separate from patrifocal ideas and current legislation in the USA and other countries, banning abortion and criminalizing attempts at abortion, emphasize the man's control of female bodies and the sperm that produce children (Hira, 2021).

Tainter (1988) in his survey of history, has argued that humans seldom respond to threats until it is too late. The fragility of civilization has been demonstrated by numerous examples (see Lamberg-Karlovsky, 1974) across time and cultures. Some, like Diamond (2004) have reclaimed the idea of environmental determinism proposed by the 19th century cultural geographers like Ratzel (1896-98), only unlike them have added a dark element of the human psyche. The self-destruction of humanity is often the theme of science fiction, yet many human societies, like those of the Indus River Valley, were quite sustainable until climate change intervened. Frankopen (2023) has summarized the varieties of human cultural exploitation of different environments and stressed the need to understand the effects on the basis of scale (size of human society) and impact as well as the scope of response necessary to achieve some accommodation given a society's institutions. These impacts have sometimes had some beneficial results for some organisms, rats, mice, crows, cockroaches and many other life forms, including those plants we have domesticated and those we call "weeds" as in Scotch Broom, have become the scavengers and opportunists of human activity. In other cases where a mutualism has arisen as with honeybirds or dolphins, Cantor, et al. (2023) note that this relationship often moves into human exploitation and aggression against the animals. We can say from both Frankopan's and Cantor, et al., that human presence acts destructively to most all cooperative ecological systems not controlled by humans. We will revisit this problem in the chapters below.

Our picture of human adaptation has been degraded due to the holocaust of colonialism which wiped off the face of the earth thousands of peoples and their cultures and records of earlier ones (Caldararo, 2003; 2022). Anthropologists over the past 200 years have tried to characterize human diversity in culture from ethnographical analysis and ethnohistorical records. But such records that have survived since the invention of writing some 5,000 years ago only touch on the latest sample of this diversity. The great body of what existed before for at least 200,000 years of the existence of our species has been lost. The record has also been distorted by what has been lost both by purposeful destruction as in the records of Native American civilizations (Caldararo, 1994) and biased description, this is seen in Diamond's (1999) telling of the peoples of Easter Island and Iceland. Neither indigenous people created a disaster, but invaders did (see Hunt, 2006). This book investigates aspects of that interaction and its origins, nature and consequences. The loss of diversity of people is a loss of human knowledge and ability, but the loss of the diversity of life is a more telling and yet a great unknown (Levin, 1999).

Leslie White (1949) points to the materialism of certain forms of Christianity as a magnifying element of environmental destruction, the type of energetic focus Weber 1958) described for Protestantism.

However, Moncrief (1970) among others suggested that religion seemed a general motivator to human environmental exploitation which all animals participate in to survive. Emmel (1977) reviews this argument and it is certain that some cultures formed social organizations and values that resulted in low or no population growth and thus sustainable conditions, while others produced institutions and values that exhausted local resources leading to either migration, conflict or collapse.

I propose here that we need a fundamental change in the social organization and institutions of present developed nations along with a new morality and focus on pleasure, curiosity and friendship with the Earth, life and our inner selves. Just as Malinowski (1927) wrote a monograph of the sexual habits of people he studied in Melanesia to contrast the stultified morality of sex and the oppression of a lack of birth control at the time, so Kenyatta (1938) critiqued European oppression of non-European culture, especially pointing out the hypocrisy of Christian missionaries who suppressed native sex customs (especially polygyny, when they, the missionaries used local women or prostitutes). This became clear to him when he studied in Europe. So today we need a new evaluation of the role of fertility in human future adaptation as well as sex and gender. Arguments of overpopulation today are as fraught with aspects of colonialism and racism as are discussions of the fate of rare animals in former colonies. Preservation of parklands once filled with "big game" hunted by white colonists have come to be last resorts for many people stressed by globalization in need of food and other resources from the Amazon to the Gobi and Kalahari. It is well established that colonial policies removed indigenous ecological practices that were sustainable, as in India and parts of Africa (Tucker, 1988; Weiskel, 1988).

While some nations have produced parks for recreation that are now being overwhelmed by the overuse of visitor presence and poorly controlled recreation (Timmons, 2019), in others, like Japan where the population is declining, pressure is reduced on the natural flora and fauna allowing for both preservation of natural ecological conditions and visitor impacts. However, what is the future of this unequal condition? Where we have a confluence of preservation and declining human population in some nations and uncontrolled growth and destruction of natural resources in others, there is an obvious Malthusian conflict building. One nation's prudent preservation policy becomes another's exploitation opportunity. Will parks created by one nation, be a source of conflict with those who lack productive land but have rising populations? There has often been this conflict in the past (Yanes, 2023).

Anthropologists of the 19th and 20th centuries concentrated much of their attention on culture change, especially of the effects of colonialism and technology to the development patterns of individuals and of social interaction (Hsu, 1972; Mair, 1969). People are born and grow up. They learn a language and come to behave as members of a group. How does this happen? It has been argued in the past that a human mind is a blank slate, a tabla blanca, upon which is written the nature of the society in which he or she is born. This model is now out of fashion for a number of reasons both due to experimental evidence and philosophical critique (Pinker, 2002). Morals and ethics are learned either by demonstration in the home or by lesson in schools. This idea of a blank slate, has come under criticism both due to experiments with twins showing some degree (usually about 50%) of inherited traits related to behaviors), and to the demise of the idea of common culture, mainly in Anthropology. If there is no common culture, no shared values and moral boundaries, then how does a society work? The fact is critiques of the theory of culture have been not only wrong, but terribly wrong and uniformed. They have been derived from ideologies and not science.

In her 1997 article critique of the concept of culture, Regna Darnell argues that late 20th century processes of "globalisation, ethnonationalism and multiculturalism all seem to create new and unique circumstances which gainsay the traditional conceptual and geographic boundaries of what anthropologists have long called culture." To make any headway in agreeing with her one needs to understand what the concepts she calls on represent. Unfortunately, "globalization" as a concept differs markedly in different publications, often seen as a continuation of colonialism. "ethnonationalism" is a term also poorly defined, yet when we look at works by Gellner (1983) or Sand (2009) we find that this process of group formation and identity is not new, and that anthropologists, like historians, have long known that social groups we have called cultures since the 19th century have origins, lives and like human bodies change over time and seem also to often disappear by either dying out or merging or morphing into new entities or "cultures." The Romans were well aware that their identity was fabricated, that they had begun as a group of refugees and bandits, homeless and beaten. Their most respected historians, like Livy and Suetonius described a number of myths and legends that purported to represent this background to an accepting public. As for multiculturalism, one can hardly go farther than Herodotus' description of the multicultural society of ancient Persia or for that matter, the Roman Empire itself. So it seems these "processes" Darnell calls upon to end the concept of culture, to "gainsay" it as contradictions to its validity, are not new or are so vague as to undermine the negation.

She claims there has never been "a single homogeneous anthropological concept of culture." She seems to be making this statement as a negative point as if dogmatic positions in the social sciences were desired or expected, while disagreements, as in the physical sciences over the origin of the universe of life were embarrassments.

Darnell triumphantly claims that much of the supposed force in destroying the concept of culture has come from an "emergent discipline of Cultural Studies." It does not seem strange to her that such a discipline established to demolish a false concept, "culture" should name itself after this false god. Nevertheless, almost everything I have read which has claimed to find fault with the concept of culture as been ignorant of the concept, wrong, or has begun by constructing a straw man of culture and then beating it with the proverbial dead horse of misunderstood theory.

I will not here provide a detailed defense of the concept of culture, it has had numerous defenders better situated that I to write them (e.g. Boggs, 2004). However, since I am going to use the concept of culture in this book, I feel it is necessary to make a few comments about critiques of the concept.

## The Nature of Method and Theory in Anthropology

During the 1960s and into the 80s and 90s a fierce struggle over criticism and the value of Anthropology divided the discipline and undermined its image in the public estimation. While earlier criticism in Anthropology had focused on methods, for example, the use of materials collected by other people, missionaries, military staff and government agencies or just tourists, the post WWII period saw a much different approach, and often drawn as personal attacks, usually based on ideology and dogma rather than methods (see examples in Manganaro, 1990. Especially that by Marty Roth). Marilyn Strathern (1990) assessed the earlier criticism, especially that of Frazer and found it basically both ignorant (lack of knowledge of major writing, out of context quotes, or misrepresentation) and simply distorted or incorrect interpretations. While Malinowski had also criticized historians like Frazer and some also took issue with anthropologists like Tylor for their use of representations of Native peoples without a critical assessment, his purpose was to introduce methods that would provide Anthropology with documented evidence and a body of literature that could be a reliable record. The method was ethnography which Boas added to and emphasized the collection of material culture and language.

Ardener (1971; 1985) entered this fray by remarking that some others had used this method as a means of aiding the colonial and racist program of othering and exoticizing non-European people. However, this also is a distortion of both the intent of Boas and Malinowski and the vast literature that was produced, one cannot control how a book is used or be responsible for crimes produced from knowledge. The structure of this work was designed to reduce or eliminate as much as possible bias; the bias seen in most of the missionary, military and political writing of government agents as well as the tendency of amateurs to idealize Natives as living surviving lives of original man, of golden eras of the past or noble savage realms. The resulting "new Anthropology," built on supposedly pure "reflexive" psychology and ideological perspectives, was to include the Native view and break down the othering of the old Anthropology. The result however, has been a production of what has been called by many of its creators as fictional writing, where the old Anthropology aimed at establishing a literature of scientific value. The new Anthropology also claimed that the old Anthropology had attempted to freeze in time Native peoples and their cultures and represent them as unchanging cartoons or caricatures. This cannot be found in the literature and is a straw man produced by the new Anthropology (read also as, "Cultural Studies") to justify its own distorted views. But the assault on what they call "Modernist Anthropology" has confused the idea of Modernism as in art, architectural and literary philosophy of the early 20th century with the ideas of late 19th and early 20th century Anthropology of which it had no part. Strathern (1990) acknowledges this problem, but the program for erasing the old and creating the new, began with a complete caricature, and to Managanaro (1990) can be summed up as, "The essays here share the contention that no anthropology is apolitical, removed from ideology and hence from the capacity to be affected by or, as crucially, to effect social formations." This statement effectively reveals the bias of the writers as they reject the program of the founders of Anthropology to create a discipline to study human kind as any other animal within a scientific and empirical context. For Managararo and similar students of Anthropology, and especially Foster, Fisher, Asad and Stocking, their political critique appears clothed in a literary analysis, like Derrida and Foucault, while they appear unbiased, yet the Marxist analysis is ever present, yet a subtext is created that is intended to be understood.

Such ideological subterfuge characterizes much of the literature critical of efforts to reduce climate change impacts. But as Friederike Otto and her colleagues (Otto, 2023) have shown, the violent nature of the consequences of climate change are difficult to communicate to the public in a way that can change national policies no matter what ideology that policy is based on, neoliberal, Marxist, corporate capitalist or religious-derived, capitalist libertarian views like the Wise Use movement. Mostly the Judeo-Christian-Islamic tradition is used by many to justify resource extraction and over-exploitation (Alper, 2022). And environmentalism, for example, which is at the foundation of groups like the Sierra Club, has abandoned the idea of population control as racist.

## References

Alper, Becka A. (2022) "Religious views on climate change," Section 4, in *How Religion Intersects with Americans' Views on the Environment,* Pew Research Center, Nov. 17, https://www.pewresearch.org/religion/2022/11/17/religious-groups-views-on-climate-change/.

Ardener, Edward W. (1971) "The New Anthropology and Its Critics," *Man*, v. 6, 449-467.

Andener, Edward W., (1985) "Social Anthropology and the decline of Modernism," In *Reason and Morality,* (Ed.) J. Overing, London, Tavistock: 47-71.

Bohannon, Cat (2023) *Eve: How the Female Body Drove 200 Million Years of Human Evoluton,* New York, Penguin/ Random House.

Boggs, James (2004) "The culture concept in context," *Current Anthropology*, v. 45, n. 2, April, https://doi.org/10.1086/381048.

Bryson, Reid A. (1973) "Drought in Sahelia: Who or what is to blame?" The Ecologist, Oct pp. 366-371.

Caldararo, Niccolo (2023) *A Mental Ethnography*, New York, Palgrave/Macmillan.

Cantor, Mauricio, Damien R. Farine, and Fabio G. Daura-Jorge, "Foraging synchrony drives resilience in human-dolphin mutualism," *PNAS*, v. 120, n. 6, https://doi.org/10.1073/pnas.2207739120.

Darnell, Regna (1997) "The anthropological concept of culture at the end of the Boasian century," *Social Analysis*, v. 4, n. 3, Nov. pp. 42-54.

de Waal Malefijt, Annemarie (1968) *Religion and Culture, An Introduction to Anthropology of Religion*, New York, Macmillan Press.

Emmel, Thomas C. (1977) *Global Perspectives on Ecology*, Palo Alto, Mayfield Publishing.

Frankopan, Peter, (2023) *The Earth Transformed: An Unfold Story*, New York, Broomsbury.

Gellner, Ernest (1983) *Nations and Nationalism,* Oxford, Blackwell.

Hira, Elizabeth (2021) "The Government has a long history of controlling women – one that never ended," *Brennan Center for Justice,* November 9th, https://www.brennancenter.org/our-work/analysis-opinion/government-has-long-history-controlling-women-one-never-ended.

Hrdy, Sarah Blaffer (1999) *Mother Nature: Maternal Instincts and How They Shape the Human Species,* New York, Ballantine Books.

Hrdy, Sarah Blaffer (1981) *The Woman That Never Evolved,* Cambridge, MA, Harvard University Press.

Hsu, Francis L.R. (1972) (ed.) *Psychological Anthropology*, New Edition, Cambridge, MA., Schenkman Publishing.

Hunt, Terry L. (2006) "Rethinking the fall of Easter Island," *American Scientist,* v. 94, n. 5, DOI: 10.1511/2006.61.412

Hutchinson, E.P. (1967) *The Population Debate: The Development of Conflicting Theories up to 1900*, Boston, Houghton Mifflin Co.

Jennings, Jesse David (1974) *Prehistory of North America,* New York, McGraw-Hill.

Johnson, Joseph D., Nathan L. White, Alain Kangsbire, and Daniel M. Abrams, (2021) "A dynamical model for the origin of anisogamy," *J. Theoretical Biology,* V. 521, 21 July, https://doi.org/10.1016/j.jtbi.2021.110669.

Kenyatta, Jomo (1938) *Facing Mt. Kenya,* London, Secker and Warburg.

Levin, Simon (1999) *Fragile Dominion: Complexity and the Commons,* New York, Perseus.

Mair, Lucy (1969) *Anthropology and Social Change*, Monographs on Social Anthropology, London, Athlone Press.

Malinowski, Bronislaw (1927) *Sex and Repression in Savage Society,* London, Routledge.

Moncrief, Lewis W. (1970) "The cultural basis for our environmental crisis," *Science*, v. 170, Oct 30, 508-512.

Otto, Friederike Otto, (2023) "This will only get worse if we don't stop burning fossil fuels," *The Financial Times,* December 6, page 15.

Pinker, Stephen (2002) *The Blank Slate: The Modern Denial of Human Nature,* New York, Penguin Books.

Rees, William E. and Mathis Wackernagel (2013) "The shoe fits, but the footprint is larger than Earth," *Plos Biology*, Nov 5th, https://doi.org/10.1371/journal.pbio.1001701.

Sample, Ian (2022) "If they could turn back time : how tech billionaires are trying to reverse the ageing process," The Guardian, 17 February.

Sand, Shlomo (2009) *The Invention of the Jewish People,* London, Verso.

Schaller, R. (1971) "Indirect sperm transfer by soil arthropods," *Annual Review of Entomology*, v. 16, 407-446.

Strathern, Marilyn (1990) "Out of context: the persuasive fictions of Anthropology," in *Modernist Anthropology*, (ed.) Marc Manganaro, Princeton, Princeton University Press: 80-132.

Timmons, Abby L. (2019) "Too much of a good thing: overcrowding at America's National Parks," *Notre Dame Law Review*, v. 94, n. 2, Article 12, 986-1018.

Tucker, Richard P. (1988) "The depletion of India's forests under British Imperialism: planters, foresters and peasants in Assam and Kerala," In *The Ends of the Earth,* (Ed.) Donald Worster, Cambridge, University of Cambridge Press, 118-140.

Wallace, Anthony F.C. (1956) "Revitalization movements," *Amer. Anth.*, v. 58, pp.254-281.

Weber, Max (1958) *The Protestant Ethic and the Spirit of Capitalism*, ed by T. Parsons, New York, Scribner's.

Weiskel, Timothy C. (1988) "Toward an Archaeology of colonialism: Elements in the ecological transformation of the Ivory Coast," In *The Ends of the Earth,* (Ed.) Donald Worster, Cambridge, University of Cambridge Press, 141-174.

Weygoldt, P. (1969) *The Biology of pseudoscorpions*, Cambridge, MA. Harvard University Press.

White, Leslie (1949) *The Science of Culture*, New York, Farrar, Straus & Young.

Yanes, Javier (2023) "History of National Parks: conservation versus expoitation," *BBVA*, 25 October, https://www.bbvaopenmind.com/en/science/bioscience/history-of-national-parks/.

Zhuang, Mingsheng, Thomas J. Colgan, Yulong Guo, Zhengyi Zhang, Fugang Liu, et al., (2023) "Unexpected worker mating and colony-founding in a superorganism," *Nature Communications,* v. 14, 5499, 1-13.

# Chapter One

# Touching in the 21$^{st}$ Century: Sexual Dimorphism, Politics and Power

**Abstract** Touching and social space have different meanings as devices of communication. Status is displayed in a variety of cultures by who touches who and how, the amount of space one person demands or can socially control is another signal of status or power. In some cultures touching and a demonstration of a lack of space regarding the presence of certain individuals is a signal of inclusion, kinship distance or amorous invitation. Generally in western cultures, especially Anglo-American, limits to touching and guarded definition of space communicate avoidance as a general means of behavior control. When social boundaries are breached in public a variety of sanctions or specific significant signals are expressed. Recently the nature of touching has taken on a specific defined political criticism and tool for sanction, yet it is not uniformly applied. This is especially true of behavioral sexual dimorphism. This contradiction is most concerning and requires analysis.

"i know that touching was and still is and always will be the true revolution." Nikki Giovanni

## Introduction

A number of elected officials, personalities and political donors have been accused of improperly touching or of aggressive and unwanted sexual attention and rape in recent years leading to the creation of the #MeToo movement. Studies of how social space is defined and how individuals of different sexes or genders can interact in public have been made by anthropologists over the past 150 years. Most notable are those by Margaret Mead (1949), Ashley Montagu (1986) and Edward Hall (1969). In more recent years, anthropologists like Jablonski (2013) have concentrated efforts on the physiology of touching as in the receptors of the skin and skin structure.

Touching is a display of power, it classifies control and expresses rank. As class and caste markings have become less visible in modern society, the distinctions between elements of the hierarchy have become more subtle. How a person dressed in the past marked clearly their status, the Roman toga is a prime example. From the Feudal period to the 19th century symbols of dress clearly acted as social dividers between the worker, farmer, clerk and lord, the merchant, business man, scientist, soldier. Aspects of dress can mark cast, class or political movement, as in 1789 with with the sans-clottes ( a reference to lower class clothing, for pants) or the term arising in the Spanish social war of 1823,"descamisado" meaning, the shirtless ones. Later it became a political term in Argentina and its Peronist movement. Although there is an earlier reference, nearly a century before Peron (O'Donnell, 2009).

We find a most interesting example of how such markers change as power changes in the modifications of female clothing at the end of the 20th century. This is a form of sexual dimorphism, and in primates we find it to be a means of classifying social structure as when males are larger than females we find, in general, polygamous breeding systems as in gorillas and some monkeys, but size does not always determine social pattern as in the solitary orang. Where there is a lack of sexual dimorphism, as in gibbons we find a form of social organization like monogamy where one male and one female form long term bonds (Hrdy, 1981). Physical differences can also be associated with behavioral differences in aggression, submission, care of young, feeding and other aspects of daily life.

In the early and middle 20th century and in the post WWII period, women were distinguished from men in general, by the dress, skirt and high heel. This was a basic form of social sexual dimorphism that showed ranking between the sexes. Some women may not have worn high heels but their shoes did differentiate them from male wear. By the end of the 1970s women's clothing was undergoing a significant change with many women wearing pants and shoes showing more comfort and practicality. This shift into the 1980s and 1990s reflected a greater role of women in economics and politics and changes in law and family relations between men and women. As power relations changed so did clothing and presently few women, especially young women, wear dresses and skirts and high heels. These types of clothing are gradually becoming more ceremonial in nature, worn to ritualized gatherings like funerals, weddings and graduations. The clothing they appear in everyday work is more like that of men and thus the sexual dimorphism between men and women has been reduced. In humans there are other aspects of sexual dimorphism that are more subtle than size or dress, for example, speech, greetings and other forms of behavioral displays, and facial architecture (Kleisner, et al., 2021). Since Darwin's 1871 book, *The Descent of Man and Selection in Relation to Sex,* debate on the relation of sexual dimorphism and human sexual organization and behavior has been significant, especially its antiquity (Larsen, 2003).

Every day males and females dress for the day to perform their gender and social role. They reproduce signs and symbols of desire, sexuality and power. These elicit responses in other individuals and often, in the appropriate context and circumstances, other people respond in culturally defined and acceptable ways. Goffman (1959) described this process and the contexts in a limited fashion, but his work is still valuable to situate the process in other ways. In most patrifocal (male oriented) cultures, as our own, women risk exposure to harassment and attack in culturally allowed means to limited culturally accepted limits during this performance of gender identity. One reason is due to the ill defined character of costume and make up regarding attraction and communication of restricting or accepting advances and changes in personal zones of contact. This has been one issue in woman's liberation movements of the last two centuries.

For men, the expression of status is associated with power and dominance. Gestures and dress are as important as in female dress and gestures in maintaining personal zones, but also these are constantly tested in some cultures, especially our own. Hsu (1980) describes and contrasts many of these based on cultural values in comparing Chinese and American society, as Hall (1959) did for a number of societies. Obviously, these methods of display and interaction have evolved in the context of survival pressures in societies of diverse types to promote reproductive success. But today they promote unneeded fertility, psychological trauma and sexual abuse. While it is generally believed that what a woman wears can effect the potential for harassment or rape, the nature of attacks on women vary so widely from desire, to anger to humiliation that it has been difficult to establish specific links (Parrot, 1988). These ritualized behaviors are reinforcing the cultural dynamics that are factors making climate change.

## Male and Female Areas, Gender Enforcement

Ideas of touching are embedded in cultural concepts of ownership and familiarity. These were described in a number of studies in the 20th century, including Margaret Mead's *Male and Female*, and several works by Edward Hall and one by Ashley Montagu. Ownership in this context refers to the idea that people are more than just members of a society or a family, but are the property of a clan or lineage and as such represent the entire group in their actions. This was exemplified in ancient literature by the single combats of heroes like Achilles and Hercules. We find the idea of houses and families or gens displayed across the Mediterranean in works by historians like Fustel de Coulanges. Usually these include the behavior of women in how the honor of the tribe is reflected in their virtue as in the famous horror of the murder of Verginia (465-449 B.C.E) by her father or that of the suicide of Lucretia (died 509 B.C.E.) of Roman history.

More diffuse are concepts of appropriate touching that might be regulated by the idea of basic status, for example of classes or castes as in European Feudalism or the Indian Caste System (Dumont, 1970). These often involve contamination, danger and purity as discussed by Mary Douglas (1966). Display of power is another aspect of touching. While a male touching a female may reflect ownership or familiarity, it can also display rank or status. When President George W. Bush gave German Chancellor Angela Merkel a surprise "shoulder rub" at a international meeting in 2006 it was deemed inappropriate from a number of viewpoints. One was that she was a married woman and he a married man, but the power relations of President of the USA to less powerful Germany was also a dynamic. It could also be interpreted as one of ownership, Germany the defeated client state of the USA is acted upon without address. The act was a demonstration of power and a means of demeaning the Chancellor both as a woman and as the leader of a lesser nation. But touching among men can have other connotations, these include power relationships, status and dominance. If a man accepts without complaint the touching of another man it indicates general submission of his relationship. If he rejects the touching it can indicate a specific challenge by the toucher both to the status of the person touched and, in certain circumstances, a threat to his masculinity if in public. This is modified by culture, but generally gives us a guideline to male on male touching, which can also indicate ownership (and sexual involvement), kinship or alliance.

Such touching of a public nature can be focused by context. When people voluntarily attend public meetings of groups of which they are members, touching is often a means of reinforcing bonds by public demonstration of kissing (for example, in Europe) shaking hands or hugging. Until very recently men shook hands as a greeting and hugged females, but this has changed in recent years with more handshaking between men and women and men to men hugging. This could be interpreted as a more egalitarian expression of world relations. Yet with the Covid-19 pandemic, less handshaking or hugging and more head-nods or fist bumps.

## Biology of Touch: Tactile Communication in Growth and Development

Montagu (1986) goes to great pains to summarize the literature on the need for touch to initiate patterns of growth and development in the normal organism. He grounds his approach in animal studies where macaque monkeys raised on model mothers, some with wire bodies and others with cloth are compared (Originally the experiments were designed and conducted by the Harlows). The monkeys raised only on wire monkeys show definite needs for more fur-like, softer tactile substrates to "cuddle." Montagu interprets a number of examples to show that development is stunted or produces pathological adaptations in behavior in later life. He uses autism syndrome as an example, though today many researchers place considerable emphasis on the role of genes, though genes act often through stimuli from the outside world and cycles of stimuli as patterns to produce normal development. The ethologist Niko Tinbergen used his substantial research in this area to produce a tactile theory of blocked development (Tinbergen, 1983). New technology has expanded this area of research and shown more specific relationships between internal and external stimulation of the developing fetus (e.g., Grossman, 2024).

Hugging females has been regarded in many public venues, especially religious meetings as in church or in political settings, to describe closeness of individuals and agreement with group principles. Such hugging broadcasts a general idea of the friendliness of the man by the relative fluidity of acceptance of the hug by women present. Recent criticism of President Joe Biden's hugging of women in public is a post-interpretation of a socially acceptable act. Such acts as demonstrations of friendliness and the humane nature of the man have also been couched in baby-kissing in campaign events. Such touching is devoid of mocking or disparaging connotations. Former Senator Al Franken's inappropriate grabbing (of Lindsey Menz) and kissing women (Leeann Tweeden) without their consent was considered a violation of space and an aggressive act, and "joking" touching of others could be interpreted as demeaning and aggressive (Zhou, et al., 2023). But such behaviors, especially with men can be acts of violence typical of bullying and part of child's play in establishing dominance (Neppi and Murray, 1997).

Donald Trump, an American President, has made a career of touching women usually with explicit sexual overtones. This presents a conundrum, as compared with Franken it seems to be a contradiction to accepted ideas of volition and space as well as what is formally acceptable adult behavior as this behavior was not consistently sanctioned or deemed problematic by many Americans. The Access Hollywood tape where President Trump boasts of touching women against their desire violates ideas of moral conduct. On tape responding to questions in a deposition, Mr. Trump claims exception that "stars" are exempt from punishment for such behavior (WSJ, 2023). How do we reconcile the difference between the punishment of former Senator Franken and the general acceptance by voters and other Americans with Mr. Trump's actions?

Some of the publicity behind the spate of accusations has come from political operatives, especially those of the rightwing, as the Israeli organization the Black Cube which was also involved in an information campaign against the Obama administration's nuclear deal with Iran.

These appeared in the media in reference to Eric Schneiderman, a piece of information or perhaps, in the currency of the 1980s, disinformation, has come to light. Opponents of the nuclear deal with Iran have accused an Obama aide, Benjamin Rhodes of fraudulently fashioning the deal to Congress. Michael Shear and Ronen Bergman in the *New York Times* (reprinted in the *San Francisco Chronicle*, May 9th 2018) report that critics of the deal hired a private investigations firm (established by former Israeli Defense Forces) Black Cube, to get dirt on those involved in the deal (Farrow, 2017). They also have been involved in gathering information that led to the accusations against Al Franken. But apparently were also hired by Harvey Weinstein, yet may have been sources on a variety of other Democratic elected officials, donors and activists. The effect on journalism was substantial in the unearthing of evil doers and exposing of sexual violations (Friedman, 2021). This is a variation of the claims ultra-right wing groups have been making against professors and teachers across the nation for supposedly suppressing their beliefs by using facts, especially attacking teachers on evolution (https://www.livescience.com/13664-attacks-mount-teaching-evolution.html and https://www.meforum.org/campus-watch/articles/2003/who-s-behind-the-attack-on-liberal-professors). Things are getting out of control and while some like Weinstein may be guilty of crimes, it is our justice system that should provide a fair trial and allow for transparent examination of the evidence. One might recall the House UnAmerican Activities Committee, the Loyalty Oaths for teacher and workers in the 1950s or the Anti-Syndicalist Laws of the turn of the century that sent thousands of union members to jail and forced them out of jobs. With the Patriot Act in place we have lost much of the protections of the Constitution, with DEA laws. Yet the point is on the loss of privacy and the nature of reports of abuse and new proposals for detentions on suspicion of use of opioids (for the public's protection) end much of the Bill of Rights appears undermined as well (https://www.sciencedaily.com/ releases/2016/12/161208090157.htm). Given ICE seizures of supposed immigrants (which has mistakenly included many citizens or stripped some of theirs: http://www.scmp.com/news/world/united-states-canada/ article/2127546/new-jersey-man-loses-us-citizenship-under-trump) and Trump's statements that critics of government policy and racism should not be citizens, the protections of citizenship are melting (https://think progress.org/trump-citizenship-flag-40df1ef1da6a/). With due process now out of vogue we are in a very dangerous landscape.

## The Magic and Liminality of the Woman's Restroom

A major political discussion broke out in the USA over the sanctity of the woman's restroom. While the euphemisms concerning where we deposit our bodily wastes are an interesting reflection of how we consider out bodies, both in private and public, the nature of the public restroom has a certain magic about it. I discuss this in some detail later on in this book. The central point here is the political assault on privacy and the nature of identity with gender at its focus. Not only do we find the abortion debate focused on the liberty and freedom of women to control their bodies, but the idea of people changing the gender under surgical and hormonal treatment is under attack. Gender enforcement is not new, most societies govern the display and performance of gender in public and private venues, but these are not always related to actual biological status. Specific behaviors are not significant in this discussion, as where hundreds of women fought in the American Civil War, or individuals took on the gender social role of the opposite sex. Rather societies usually hold families to be reproductive units and their products, children are a focus for gender roles. I will address this further in a later chapter.

## References

Dumont, Louis, *Homo Hierarchicus,* Chicago, University of Chicago Press, 1970.

Farrow, Ronan, "Israeli Operatives Who Aided Harvey Weinstein Collected Information on Former Obama Administration Officials," *The New Yorker*, May 6, 2018.

Friedman, Lynn (2021) "The impact of the Me Too Movement's Journalism," *Yale Journal of International Affairs,* May 15, https://www.yalejournal.org/publications/the-impact-of-the-me-too-movements-journalism.

Goffman, Erving (1959) *The Presentation of Self in Every day Life*, London, Penguin.

Grossman, Tobias (2024) "Brain origins of early social behavior," *Behav. Brain Sci.,* Jun 27:47:e128, doi: 10.1017/S0140525X23003151.

Hall, Edward, (1969) *The Hidden Dimension*, Garden City, Anchor Books.

Hall, Edward (1959) *The Silent Language,* New York, Fawcett Books.

Hrdy, Sarah Blaffer (1981) *The Woman That Never Evolved,* Cambridge (MA), Harvard University Press.

Hsu, Francs L.K. (1980) *Americans and Chinese: Passages to Differences,* 3rd edition, Honolulu, University of Hawaii Press.

Jablonski, Nina (2013) *Skin: A Natural History*, Berkeley, The University of California Press.

Kleisner, Karel, Petr Turecek, S. Craig Roberts, Jan Havlicek, Jaroslava Varella Valentova, et al., (2021) March 16, *Scientific Reports,* 11, 5978, https://www.nature.com/articles/s41598-021-85402-3.

Larsen, Clark Spencer (2003) "Equality for the sexes in human evolution? Early hominid sexual dimorphism and implications for mating systems and social behavior," *PNAS*, August 5, 100 (16): 9103-9104.

Mead, Margaret, *Male and Female,* New York, HarperCollins, 1949.

Tinbergen, Niko and E. Tinbergen (1983) *Autistic Children: New Hope for A Cure,* London, Allen and Unwin.

Montagu, Ashley (1986) *Touching, The Human Significance of the Skin,* New York, Harper & Row.

Neppi, Tricia and Ann D. Murray (1997) "Social Dominance and play patterns among preschoolers," *Sex Roles*, v. 36, n. 5-6, DOI: 10.1007/BF02766654

O'Donnell, Pacho *(2009). Juan Manuel de Rosas, el maldito de la historia oficial.* Buenos Aires: Grupo Editorial Norma.

Parrot, Andrea (1988) *Coping with Date Rape and Acquaintance Rape,* New York, Rosen Publishing.

Wall Street Journal, May 2023, https://www.wsj.com/video/in-deposition-trump-says-historically-stars-can-grab-women-by-genitals/9925F3D8-9B68-45F9-9DC1-83467D65F42C.

Zhou, Zhuang, Xiaogang Zhou, Gang Shen, Ahmad Zamri Khairani and Jamalsafri Saibon, (2023) *Psychol. Res. Behav. Manag.* V. 16: 5041-5051.

# Chapter Two

# Hair, Editing the Body and the Identity of Membership in Human Evolution: Monocultures and Pandemics, Why Diversity Matters in Climate Change

Now and then books and articles appear that treat certain features of humans as special, distinct from any other animal. Freakish is almost implied as if human existence was an abnormality on the planet. One such book was Jared Diamond's *Why Sex is Fun* (1997) that labored to demonstrate that human sex was so different from all other sex among animals that it was almost outrageous. This could only be accomplished by the tortured stereotyping of human sexuality, mainly that is, stereotyped ideas of middle class American and European sex. For example, he claimed that humans were the only animal that had sex in private. This would be a surprise to the thousands of Americans who are charged with sex in public places each year (Moore, 2015). A similar attempt was *The Naked Ape* (1967) by Desmond Morris, which so astounded Ashley Montagu that he titled his critique, "The naked Englishman." What struck Montagu as strange was that Morris was a respected zoologist who seemed to think that all there was to talk about regarding human body hair could be found on the streets and heath of that generally foggy nation. Of course, Montagu had written an article a few years before where he claimed that, "Man is the only hairless primate,..." (Montagu, 1964). Thus Montagu had lent himself a degree of license by the title, however, which used the words, "near hairlessness." Nevertheless, it is only the sensational headlines of such claims that reach the popular press, the corrections seldom do.

In recent years many of the old distinctions that seemed to separate humans from other animals have either fallen away or been severely limited by new research. Humans are no longer regarded as the only tool-users, many primates use tools routinely, capuchins with great dexterity. Nor is man the only toolmaker, chimpanzees, ravens and crows are known to produce tools. This complicates discussions of the changes in cognition and physiology which seemed to characterize human evolutionary success. Beavers make houses, dams and change the environment. We find complex societies in ants that also have agriculture and we now know they apply fertilizer (Moore, 1996). Ants fight wars and take prisoners making them slaves. Even language seems undeniably present by most definitions in dolphins and certainly in bees (see a discussion of these comparisons in Caldararo, 2017).

The more we learn about the natural world the less "freakish" Homo sapiens appears. The big brain is not that big compared to the dolphin in brain to body ratio and when we exercise the scale of brain size we find that there is a relationship in the animal world between increasing brain size, larger neurons and lower density and metabolism (Jerison, 1973; Aiello, & Wheeler, 1995). Thus the "big brain" seems even smaller than ever today as we become aware of many of its complexities (Striedter, 2005). Yet in another case, it was a surprise to read recently that some scientists had found that human hair was quite, well, unusual (Neufeld & Conroy, 2004). While Aristotle had noted that humans were featherless bipeds (he did not know of penguins) one imagines that contemporary sages would be more informed about the diversity of traits in the animal kingdom.

## Hairiness, Membership and Civilization

The article by Arthur Neufeld and Glenn Conroy shocked me as I was familiar with Professor Conroy's work as a biological anthropologist. Neufeld and Conroy (2004) claim that human head hair is not fur. They separate the idea that humans lost their body hair from this discussion sparing themselves the embarrassment of Morris. Explanations of body hair reduction in humans are numerous as Neufeld and Conroy note, and research by Alan Rogers (Rogers, et al, 2004) that one of the genes affecting skin color is involved in hair reduction would seem to support the thermoregulation theory. Hair holds in moisture producing an insulating affect that also reduces moisture lost. This is ok as long as one is not running or very active in the middle of the day in a hot and dry environment as East Africa was about a million years ago. Claims that hair reduction occurred in response to parasites belies the fact that humans are still plagued with them nonetheless and does not explain the density of head hair in most people. "Most" is the important word here for, as in the case of Morris, not all people have the kind of hair nor the rate of growth asserted by Neufeld and Conroy. Thierry's (2005) and Jolly's (2005) championing of head hair as a communicative device is of interest and reflects some of Darwin's musings (1871) though others (Synnott, 1987; Leach, 1958; Hallpike, 1978) have produced comprehensive analyses of the subject.

Neufeld and Conroy believe that human head hair is characterized by long hair which grows continuously throughout life. Thick head hair would seem to argue against it being a thermo-regulating device (but see Falk, 1990). This would be especially true for dark hair, but then we know that hair on the head is infused with moisture from the scalp during activity or in hot temperatures and forms an insulation layer reducing the effects of heat gain. The rest of the body gives off heat by evaporation also reducing heat buildup.

The clearest problem with Neufeld and Conroy's theory is that most of humanity does not have one uniform hair type. Second, many hair types do not grow in straight, continuous fibers like the terminal hair of Europeans and many Asians. There are generally 4 types of hair, lanugo or infant hair, primary terminal (children's hair), secondary terminal (adult hair) and vellus hair. Hair fiber length and shape is governed by a number of factors (Ryder, 1958) including diet which affects both the sebaceous glands and nature of hair production as well as the hormones which also vary by sex and population as well as activity. Most humans with straight hair produce a head hair growth of about 16cm per year to a maximum of 3 feet. However, hair growth in the mid to late twenties undergoes hormonal changes that slow growth for head hair which begin to grow shorter and finer (Robbins, 1994). Neufeld & Conroy refer to recent experiments showing that terminal hairs transplanted to the legs grow at a slower rate than on the head. This confirms knowledge that terminal hairs are found naturally on the legs (1994) and that proliferative tissue behaves differently depending on its location and influence of local hormonal and other effects. This was reported by Garn in 1948 (Garn, 1951). It has been known for more than 50 years from skin grafts that they produce hair shafts characteristic of the donor site (Garn, 1951).

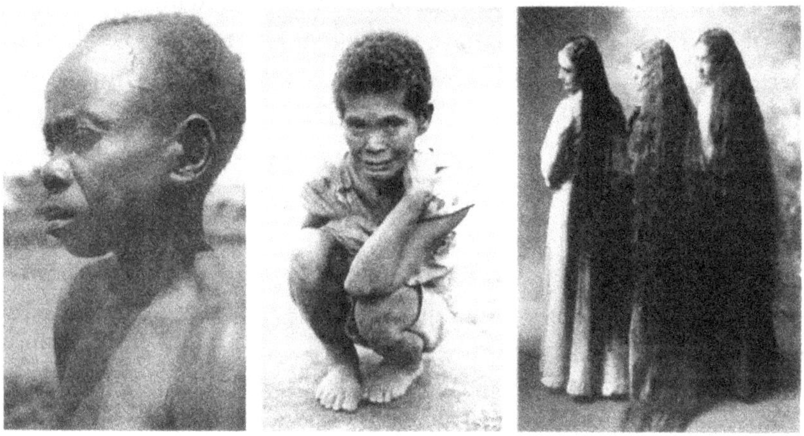

**Figure 2 Some Hair Variation Found in Human Populations**

There are unusual examples of continuously growing hair to unusual lengths that is the result of peculiar conditions of diet and lifestyle (sedentism) and hormonal abnormality. Some normal population variations are shown in the illustration, figure 2 from my article on hair (Caldararo, 2005). If we were only to focus our attention on humans, however, we would be doing a disservice to the study of hair. Like any proliferating tissue, hair has a relationship to development, diet and maturation (Price, 1958; Bradfield and Bailey, 1967). Hair pigmentation differs between individuals and groups (Hausman, 1925; Hausman, 1928). A study by Myers & Hamilton (1951) found a significant difference in hair growth in males and females, with hair growth faster in males. Seasonal growth rates have been quantified (Randell & Ebling, 1991). But the nature of continuously growing hair is found in some other animals, namely a few that have been selected by humans for this trait. Thus hair growth is partly a product of domestication. The idea that humans are a domesticated animal was first suggested by John Bachman (1850).

We find in sheep continuous growth of hair (Ryder, 1958) in domesticated varieties and this growth varies from other members of the same genus. This is especially true with the Merino sheep that has developed persistently growing hair follicles (Norback, 1951). Horse mane and tail also seems to grow continuously (Roser, 2005). This is to be expected since we know that hair growth varies significantly in different areas of the body of a variety of animals, and among different human populations (Hamilton, 1958; Saitoh, et al., 1967) as well as within human groups. The factors, developmental and genetic, affecting the ontology of hair growth are not clearly understood.

## Editing the Body

How do we recognize someone as friend or foe? In the past membership in a group depended on both appearance and behavior. We have evidence of tattoos, scarification and mutilation in the form of figurines, cave drawings and effigies from as early as 200,000 or 300,000 years before the present in the Venus of Berekhat Ram (Marshack, 1997), but at least as old as 40,000 B.P. (McDermott, 2021). In many cultures we find groups, even with histories of ancestry or mythological unity, wearing different colors or headdresses, while the same or similar types of differences can be applied to the genders in any society, even in the Amazon where clothing is minimal, differences appear in both adornment (including body painting) and hairstyle (Meggers, 1996). Below are some of the tattoos that appeared on the skin of the man discovered in ice in Tyrol and dubbed "the iceman" or "Otzi." His death has been dated to about 5,300 B.P. and he was a 61 year old man. He may have died of a fall, or a stroke yet he was encased in ice for over 5,000 years and his body well preserved by the conditions (Geggei, 2015). Understanding the tatoos' meanings is rather difficult without knowing his culture, but we do know something about how he lived and what he ate. Joyce (2008) has summarized much of the archaeological work that gives us a slight idea of how the body was used in the past. As she notes, most of human history is lost to us, we have only objects and residues of the life people led.

As in the case of the tribes of the Amazon, Otzi's people had adapted to certain conditions of life, plants and animals that could be used to make tools, shelter, clothing and as food. The specific adaptations of their culture had come about over a period of thousands of years and we could say they were sustainable, both for the Amazon and the Alps. Today's way of life is partly the product of a great disaster, the Black Plague appearing in1346. Life was beginning to be redefined at the time due to new contacts with different zones of life in the world, due mainly to trade, climate and the movement of large numbers of people across vast territories of the Eurasian landmass. This movement and development coincided with a climate change that affected the distribution of rats and their fleas that carried the bacteria that produced the Bubonic Plague or Black Death. The climate change also resulted in both the Mongol onslaught on Europe and the Middle East and the Islamic conquest of the Judeo-Christian Holy Land. Both disease and movements of people and conquest created a need for new information or old information and new ways of using it. This produced the Renaissance, the voyages of discovery and the Industrial Revolution. So our present way of life, is the product of new thinking based on necessity but it is only some 700 years old as a social adaptation (Caldararo, 2012).

Yet our earliest written histories, supported by archaeology and genetics, place the beginning of the first experiment in civilization in new lands, created from the warming at the end of the last glacial period. Bottero (1992) after decades of study unwraps the origins of Sumerian life in the necessity to cooperate in controlling the changing rivers of Mesopotamia.

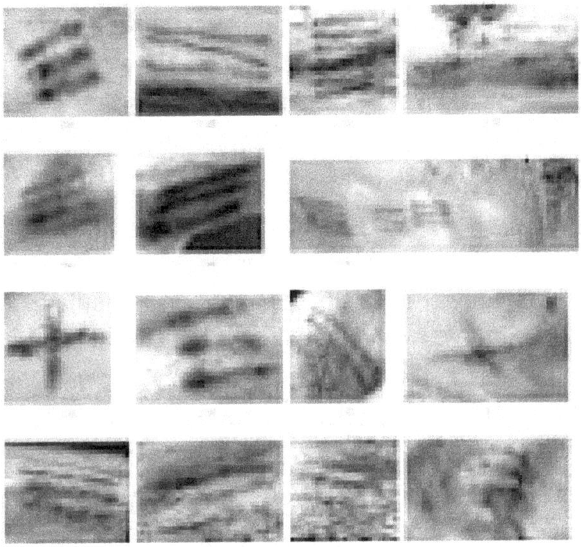

**Figure 3 Otzi's Prehistoric Tattoos**

Many people in history have changed their bodies, some adding images, others removing hair or shaping it, while some have reduced or eliminated parts of the body. Covering the body is typical of *Homo* species as long ago as the appearance of the awl in the refuse of *Homo erectus* in the temperate zone of the Earth.

## Species Recognition, Mate Competition in Gesture and Elaboration

While it would seem likely that body modifications as in tattoos, hair varieties and other forms we have mentioned should function as important attractors and discriminators of potential mates, the mate hunt is more often associated with species and culturally specific signals of males and females. As biological males and females (cis genders) may act as other genders for a variety of purposes (pleasure, wealth, celebrity, etc.), this cross gender performance is not unusual in nature and is found in a number of animals.

Some examples of this can be associated with a subordinate male or female gaining access to mates that they would otherwise be unable to forcefully compete for, while in other cases it seems a means of population control and has been identified with forms of altruism (Trivers,1985). While the fact of the use of this tactic to engage in a successful coitus can be seen as a process of increased diversity, the act of sexual altruism is often linked with increased fitness of relatives. A third possibility has now and then been suggested, that of population control, but has seldom been explored, especially for humans. I mentioned this in the *AIDS & Society Newsletter* in 1990 when attacks on sexual freedom were rife in the wake of the AIDS pandemic (Caldararo, 1991). Such ideas have usually been associated with studies of overpopulation, both in human societies and domesticated animals. Calhoun's studies, popularized in a 1962 *Scientific American* article were restudied by Hammock in 1971. The results with rodents seemed to be verified that one of the mechanisms of population control in certain conditions was homosexual behavior. The idea has been undermined by recent studies that show homosexuals, male or female, often desire children and either have them with cis partners, adopt or use IVF or surrogates (Wilson and Bouton, 2024). We will explore this more later, but this approach is not limited to mammals, we find in insects, specially the bat parasite Afrocimex, an adaptation where males appear to be transvestites, bearing pseudo-female paragenital structures which invite homosexual attention. Some biologists (e.g., Hinton, 1964) felt that the resulting sodomies were a means of food sharing, but also can be seen as sperm stealing by males who then impregnate females with their own sperm (Sivinski, 1980).

## The Sacred Wonder of the Woman's Restroom

A current hysteria as gathered around two aspects of transsexual existence. One is the "Drag Show" which is often a performance done by a man in a dress that mimics the stereotype behavior of a women. Phyllis Diller once said she was a female drag queen. Generally many men who attended such show in the past, like strip tease performances in men's clubs, interpreted them as satires of women's roles or as sexually stimulating entertainment.

In the 1950s and 1960s such drag shows were considered by most gay people I knew as demeaning to gay men, as it was thought to be a caricature of all gay men, and to gay women and to many straight women they were seen as also demeaning for many of the same kinds of interpretation. “Drag shows” were called Female Impersonation shows in polite society in the pre-WWII period and such mimicking of women has taken a number of forms in traditional societies, from shaman to Halloween costuming to various holidays of cross-dressing acts associated with libertine activities (Newton, 1972). In the last 2 decades they have had a tremendous resurgence both because they were used as a means of creating sensational news during the AIDS epidemic. This was especially emphasized by the Sisters of Perpetual Indulgence in San Francisco. And due to a reinterpretation by cultural philosophers of the gay movement who have given the performances a new definition as an art form of gay culture (Beckner, 2017). For Women and cis gay women also, this still is a tenacious subject (Murphy, 2014).

## References

Aiello, Leslie and Peter Wheeler (1995) "The Expensive-Tissue Hypothesis," Curent Anthropology, v. 36, n. 2, 199-221.

Bachman, John (1850) The Doctrine of the Unity of the Human Race Examined on the Priniples of Pure Science, Charleston, C.G. Canning.

Beckner, Asher (2017): "Recignize the purpose and cultural significanceof drag shows," *The Red & Black,* November 6, https://www.redandblack.com/opinion/opinion-recognize-the-purpose-and-cultural-significance-of-drag-shows/article_c58ba7ec-c282-11e7-9b94-7bddabab83bc.html.

Bottero, Jean (1992) *Mesopotamia: Writing, Reasoning and the Gods,* Chicago, University of Chicago Press.

Bradfield, R.B. & Bailey, M.A., (1967) "Hair root response to protein undernutrition," in Montagna, W. & Dobson, R.L., eds., *Hair Growth*, Advances in Biology of Skin, v. 9:109-119.

Caldararo, Niccolo (2017) *Big Brains and the Human Superorganism,* Lanham, Rowman & Littlefield.

Caldararo, Niccolo (2013) *Some Evolutionary Aspects of Disease Avoidance,* Saarbrucken, Scholar's Press.

Caldararo, Niccolo (2005) "Hair, Human evolution, and the idea of human uniquesness," *Evol. Anthro.,* v. 14, n. 4, 132-133.

Caldararo, Niccolo "Commentary and Opinion," *AIDS & Society,* January / February, v. 2, n.2, 1991:3.

Calhoun, J. B. (1962) "Population density and social pathology," *Scientific American*, v. 206, n.2:139-148.

Darwin, Charles, (1871) *The Descent of Man and Selection in Relation to Sex*, Republished, 1981, Princeton, Princeton University Press.

Diamond, Jared, (1997) *Why Sex is Fun*, Basic Books, N.Y.

Falk, Dean (1990) "Brain evolution in Homo: the 'radiator' theory," *Behavioral and Brain Sciences,* v. 13, 333-381.

Garn, Stanley Marion, "Types and distribution of the hair in man," in J.B. Hamilton, ed., *Growth, Replacement and Types of Hair,* Ann. N.Y. Acad. Sci., v. 53, 1951:498-507.

Geggei, Laura (2015) "Researchers just found a new tattoo on 5,300-year-old Otzi," Insider, *LiveScience*, January 30, https://www.businessinsider.com/new-tattoo-found-on-otzi-the-iceman-2015-1.

Hallpike, Christopher R. (1978) "Social Hair," in *The Body Reader: Social Aspects of the Human Body*, ed. Ted Polhemus (New York: Pantheon Books, 134-46.

Hamilton, J.B., (1958) "Age, sex and genetic factors in the regulation of hair growth in man: a comparison of Caucasian and Japanese populations," in *The Biology of Hair Growth,* eds., W. Montagna & R.A. Ellis, Academic Press, N.Y.:400-434.

Hammock, James Robert, (1971) "Behavioral changes due to overpopulation in Mice," *Dissertations and Theses,* Portland State University, 7/26/.

Hausman, L. A., (1928) "The pigment granules of human head hair: a comparative racial study," *Amer. J. Phys. Anth.,* v. 12, n. 2:273-283.

Hausman, L.A., (1925) "A comparative racial study of the structural elements of human head-hair," *The American Naturalist,* v. 59, n. 665,:529-538.

Heufeld, A.H. & Conroy, G.C., (2004) "Human head hair is not fur," *Evolutionary Anthropology,* v. 13:89.

Hinton, H. E. (1964) "Sperm transfer in insects and the evolution of hemocoelic insemination," in K.C. Highman, Ed., *Insect Reproduction Symposium*, No. 2, Royal Ent. Soc., London, pp. 96-107.

Jerison, Harry J., (1973) *Evolution of the Brain and Intelligence*, N.Y. Academic Press.

Jolly, Alison, (2005) "Hair signals," *Evolutionary Anthropology*, v. 14, n. 1,: 5.

Joyce, Rosemary A. (2008) *Ancient Bodies, Ancient Lives: Sex, Gender and Archaeology,* New York, Thames and Hudson.

Leach, Edmund, (1958) "Magical Hair," *Journal of the Royal Anthropological Institute* 88: 147-64.

McDermott, Ann (2021) "What was the first 'art'? How would we know?" *PNAS*, v. 118, n. 44, https://doi.org/10.1073/pnas.2117561118.

Marshack, Alexander (1997). "The Berekhat Ram figurine: a late Acheulian carving from the Middle East." *Antiquity*. 71 (272): 327–337.

Mazziotta, J.C., et al., (1995) "A probabilistic atlas of the human brain: theory and rationale for its development," *Neuroimage*, v. 2:89-101.

Meggers, Betty (1996) *Amazonia: Man and Culture in a Counterfeit Paradise,* Revised edition, Washington, D.C. Smithsonian Institution Press (Originally published in 1971).

Morris, Desmond, (1967) *The Naked Ape*, Crown Books, N.Y.

Montagu, Ashley, (1964) "Natural Selection and man's relative hairlessness," *Journal of the American Medical Association,* v. 187, Feb. 1:356-7.

Moore, Peter (2015) "One third admit they have had sex in public," *Yougov.,* May 13.

Moore, Peter D., (1996) "Invertebrates and mycophagy," *Nature*, v. 381: 372-3.

Murphy, Meghan (2014) "Why has drag escaped critique from feminists and the LGBTQ community?" *Feminist Current,* April 25, https://www.feministcurrent.com/2014/04/25/why-has-drag-escaped-critique-from-feminists-and-the-lgbtq-community/.

Myers, R.J. & Hamilton, J.B., (1951) "Regeneration and rate of growth of hairs in Man," in Hamilton, J.B. ed., *Growth, Replacement and Types of Hair*, Ann. N.Y. Acad. Sci., v. 53:562-8.

Neufeld, A.H. and Conroy (2004) "Human head hair is not fur," *Evol. Anthro.,* v. 13, 89.

Newton, Ester (1972) *Mother Camp: Female Impersonators in America,* Englewood Cliffs, Prentiss-Hall.

Norback, Charles, (1951) "Morphology and phylogeny of hair," in J.B. Hamilton, ed., *Growth Replacement and Types of Hair*, Annals of the New York Academy of Sciences, v. 53:476-492.

Price, Dorothy, (1958) *Dynamics of Proliferating Tissue,* University Of Chicago Press, Chicago.

Randall, V.A. & Ebling, F.J., (1991) "Seasonal changes in human hair growth," *Br.J. Dermatol.,* v. 124, n. 2:146-151.

Robbins, C.R., (1994) *Chemical and Physical Behavior of Human Hair,* 3rd Ed., Springer-Verlag, N.Y.

Rogers, Alan R., David Litis and Stephen Wooding (2004) "Genetic variation at the MC1R locus and he time since loss of human body hair," Current Anthropology v. 45, n. 1, 105-108.

Roser, Dr. Janet F., (2005) Dept.of Animal Science, University of California, Davis, Personal Communication.

Ryder, M.L., (1958) "Nutritional factors influencing hair and wool growth," in *The Biology of Hair Growth*, eds., W. Montagna & R.A. Ellis, Academic Press, N.Y.

Saitoh, Masaji, Uzuka, M. & Sakamoto, M., Kobori, T., (1967) "Rate of hair growth," in W. Montagna & R.L. Dobson, eds., *Hair Growth,* Advances in Biology of Skin, v. 9: 183-201.

Sivinski, John, (1980) "Selection and insect sperm," *The Florida Entomologist,* Mar. v. 63, n. 1, pp. 99-111.

Striedter, Georg F. (2005) *Principles of Brain Evolution,* Sunderland, Sinauer Associates, INC.

Synnott, A., (1987) "Shame and glory: a sociology of hair," *The British Journal of Sociology*, v. 38, n. 3, Sept.:381-413.

Thierry, B., "Hair grows to be cut," *Evolutionary Anthropology*, v. 14, n.1, 2005:5.

Wilson, Bianca and LaurenJ. A. Bouton (2024) LGBTQ Parenting in the US, *School of Law, Williams Institute*, UCLA, July, https://williamsinstitute.law.ucla.edu/publications/lgbt-parenting-us/.

# Chapter Three

## The Cockettes and Gender Variation, Translation of Sex and Fun in the 1970s

Gilbert Herdt (1981) describes the manner in which the Sambia of Papua New Guinea engaged in initiation rites and purification that involve men and boys performing acts on male genitals that would, in America or in much of the world, be seen as homosexual or homoerotic. But it seems the Sambia have no concept of genital gratification or sexuality as would be considered elsewhere in these acts. This creates a dilemma, if an act does not result the end product as expected in most similar cultural contexts, how do we regard it? The same problem develops when we compare ideas of gender and marriage in a variety of cultures. If gender is not discontinuous in recognition, if the possession of certain organs does not define a role of male or female is it legitimate to assign one? One can argue that this is absurd, that women, with certain ritual exceptions (e.g., Catholic an Buddhist nuns, Albanian Honored women; Vestal Virgins) have sex with men and have children. Men can have sex with other men but do not have children. Excepting also cases of infertility, even a large number of apparently homosexual and bisexual women have pregnancies and give birth. The entire basis for the idea of a continuous spectrum of genders and a rejection of the traditional binary perception seems false from this perspective.

Usually we assign specific genders to social roles. Parenting and the provision of care and nourishment is generally identified as a female role in human societies. In non-human primates this is also a general distinction with the exception of some New World Monkeys, the Tamarins, especially. In some mammals and other animals, amphibians, insects and spiders, males may be the only care and protection provider (Trivers, 1985).

The process of attraction is quite varied among mammals, while smell is of some significance, visual stimuli are prominent. In humans we find artificial modifications of the body including tattoos, clothes, colors and makeup. In some cultures face painting is done in a religious context as in rites of passage, in others it has a unique role in sexual attraction or sexual dimorphism (Magee, 2012). We find in our own culture that women have often worn makeup as a general form of social identity with types and varying amounts of color associated with specific contexts, formal, religious, work and sex. In a few cultures men wear makeup, as among the Wodaabe to attract female interest but generally only in specific times and places.

**Figure 4 Wodaabe Men From BBC Film**

## Origins

Much has been written about the performance group the Cockettes. Much of it sensational, mainly because the Cockettes meant to be unique, outrageous and creative. Probably the most distorted rendition of who the Cockettes were is the film by David Weissman. But in my conversations with him he did seem conflicted concerning who and what the Cockettes were. Probably the term, "genderfuck" originated in reference to the group. This was partly due to the fact that some of the Cockettes were very political and were dedicated to make sexuality free of gender stereotypes and cultural morality. It was also because some of them, like Hibiscus (George Edgerly Harris, III), were so completely determined to shock by their own desire to act in an unrestrained fashion, that they always appeared dressed in confrontational manner. That is not to say that all the Cockettes or their work was confrontational, but that was often how they were perceived.

Perception is often described by magicians as 90% of any trick (to approximate the image in behavior and to create distractions at the same time), and among the transvestites and transsexuals of the 1960s who made up part of the community from which the Cockettes emerged, to act like a woman was often a skewed and confused performance, as if, on the one hand, they were attempting a homage and yet on the other, creating something that appeared as a satire. A small subset of cross-dressers/transsexuals genuinely, in my estimation, and in their own words and acts did attempt to "be women." In these the study of mannerisms, dress and style as well as voice all mimicked actual behaviors of cis-women and a few were "passable" or were seen and responded to as women. These people often were retiring and appeared shy, avoiding contact as if it would reveal their true nature. One never spoke in public but would pretend to be mute and carried a small tablet that he wrote on and communicated with others by, he tried never to laugh in front of others and was so embarrassed once then he sneezed that he left and went home from a party. Her main strategy was to be little seen, rather present but somewhat invisible. The general public expected such "performers" to skirt danger by arousing in heterosexual men murderous attack. The reality seems to be less violent and more pathetic given the merciless, and savage responses that took place that I heard of and have been generally reported. A terrible example is that of Gwen Araujo who was dressed as a woman, passed well, but attended a party in 2002 where almost all the people were heterosexual. She engaged a sex with several men and when found out (by a female party-goer) she was beaten and strangled to death (Fraley, 2016). Many transsexuals and transvestites spoke of this kind of violence. They feared the outcome, but they were always in danger of discovery. I have to say that I never saw this, and I was never attacked, or harassed when in a dress. And I cannot say it was because San Francisco was a unique environment as I often traveled to other parts of California and once in New York as a woman with no problems and flew in planes. Friends thought that was because I passed well, but since I was like the person above, somewhat invisible and quiet (as an observer should be) that is unknowable.

## Milieu of Social Performances and Plasticity of Identities

One might argue that in any period of social experimentation, the Renaissance, Roaring Twenties, the 1960s or times of great change in conception of social order as in the origins of Buddhism, appearance of Confucius or Christ, there takes place an upheaval in the idea of what is expected of people. I would say that with the Cockettes, in the milieu of the time, a curtain had risen on what was possible and permitted. If you can imagine getting up in the morning and deciding to dress as a male or female, as a 19th or 17th century person, as a donkey or panda and then take to the streets with friends doing much the same thing, you might come close to the tenor of the bubble of time the Cockettes lived in. Is it freedom or insanity to live by whim and reinvent yourself constantly? If the measure of the day is the fun you can have, the invention in food, art or sex you can experience, then you are close to the aroma of the Cockettes. This sometimes seems to verge into the realm of masking (Crumrine and Halpin, 1983) but without the traditional custom and community. While at other times it appears like, though more limited, the participation of people in various cosplay communities (Travers, 2024).

It could be regarded as a mania, but a mania of discovery and invention. As everyday would be spent watching movies, TV serials, commercials, going to second hand stores for clothes, army surplus stores, or exploring the museums, parks, libraries and streets of the city for information and inspiration. All this added up to not only a constant journey in search of the stuff of theater, but of the mystery of life and meaning. Discussions around the breakfast table or dinner, in hallways or bedrooms, were often flooded with concerns of shows, plays, skits, songs and costumes. It was in many ways like a school or cloister where the students or inmates swam in a sea of questions and propositions, attempting to mix together loose ends into a workable performance, while on the run amid an ant hill whose rushing members were like gas atoms, yet now and then one of these "atoms" would join the group by accident.

These were often forays into the forbidden, not just drugs but into places and actions. For the Cockettes to go to a church or museum often caused some distinct reaction from those regulars to such cultural institutions and their caretakers. It was impossible to ignore a troupe of men and women dressed in various outfits from different times or roles and mixed gender signals. Danger was often present, though I cannot recall anytime when violence actually took place. A close call occurred when a motorcycle friend of one of the Cockettes invited them to Fort Bragg, a town that is quite a tourist attraction north of San Francisco in Mendocino County. Some 10 or 15 outlandishly dressed Cockettes stuffed into two vehicles and drove to the town. Due to a mistaken address and directions, the group was forced to drive around town and ask people for specific landmarks. In the process of this, the bearded, dress-wearing, high-heeled leaders decked in garish makeup were often misunderstood and their intention met with disbelief. One gruff man of indeterminate age threated "queers" with violence and a couple of middle aged women informed them they were calling the police or wishing them to hell.

Hibiscus (George) would routinely say, "Just act like you are real, demand a presence and people will see what you want to be." I think the name, "George," often associated by Americans like "Edward" with kings had something to do with this attitude. But then there is an aspect of the name "George," that seems to assert both regal nature as well as transformation, as in Boy George the singer. At the time of the Cockettes, one friend of theirs, Miss George who was in the entourage of the actress, Devine, had this quality. Miss George had a presence and a transformative aspect. In anthropology we call it being liminal, or between the normal and the sacred. Miss George looked like a woman, both in body shape, stature and demeanor. Part of this was attitude as Hibiscus would argue, but also there was a skill that he/she had developed since childhood to project the persona of a female. Miss George was a biological male, but appeared everywhere in woman's attire and no one ever, as far as I knew, could, as they said then, "clock her." That is, identify her as a man. She/he had not undergone sexual reassignment surgeries.

Impressionists often argue that the ability to confuse an audience and produce belief, as in the case of faith healers, that their actions are genuine is dependent on two main factors: the degree they can attain to mimic reality and by approximation to the original and the creation of expectation (Goffman, 1959; Randi, 1993). Suggestion has been argued to be the foundation of hypnosis and Rivers (1920) drew a convincing evolutionary basis for how powerful suggestion could be. Lewis (1977) showed that it could under certain circumstances result in death, referred to as psychogenic death. So the point to this portrait is that the creation of impressions among the Cockettes was bound up in personal desire to achieve certain states and the process often seemed magical. In fact, reviews of their first shows in San Francisco were repeatedly called magical, fantastic and psychedelic. Of course, this was the time of psychedelics (see my book for more detail on this, Caldararo, 2022).

Many a beautiful afternoon I sat with Link, one of the Cockettes who often produced the book, or story for the shows. He would discuss movies, novels, peoples' lives he had been told, dreams and hallucinations, and from them he would devise a script for a show. Usually these carved up and repurposed some 30s movie or novel with bizarre added features as I have recalled. As the afternoon progressed to night, over meals, marijuana, peyote or LSD, his bed would be covered with sheets of paper, some cut into pieces and glued together eventually into a corpus for production. This would be discussed with others, sometimes in the house several of the Cockettes lived in as a communal lifestyle where bedrooms had been cut up with lofts set in so that as many of 10 or 12 people could live in one flat. I did not live in the first Cockette houses, but later moved into one that was a storefront on Valencia street before the New York Trip and shows. The storefront was used as a rehearsal studio and classroom as well as a place to manufacture props, some costumes and hold meetings. It also hosted dinners and parties as well as an after show and club hangout.

I used to sleep in one of the window boxes of the storefront where merchants usually displayed their goods. Another person, a friend of Scrumbly's slept in the opposite one, each facing the street with big picture windows on either side of the front door. These windows were covered with either posters of shows, blank or colored paper or painted. It should have been noisy, but in retrospect it seemed both quiet and private. Several sheets of material that had once been scrims for showpieces, hung separating the inner space from the main room. I worked at a job during the day, but there were still often classes in the evening or rehearsals in the space. These were not just for Cockette shows but for private lessons and other shows and performances as well. A few people lived in a balcony, and Scrumbly and Sweet Pam occupied the main rooms above a kitchen area and behind that was a wardrobe room with people's clothes and discarded costumes that were often being cannibalized for new ones. Below that was a room used for making props and communal sleeping, with mattresses from one wall to the other.

People were not especially promiscuous or monogamous, heterosexual or homosexual, some were decidedly unisexual, some non-sexual (as much as could be discovered) others were what is often today called polyamorous. But they were still human. One night I came home after work, ate and went to bed around 11pm after studying, as I was working on my Master's Degree. Around 3 or 4 in the morning a group of people came home from the clubs and had brought a few stragglers who needed a place to sleep as they were too drunk, drugged or tired to go to their respective abodes. I was fast asleep, but I was told the next day that after all the couches and spaces in the communal room were taken, there was still one person left. She was about 22 years old and from Georgia, having just arrived that day and had a suitcase. She had been reluctant to join the group in the communal room and forlornly asked if there was extra bedding or a sleeping bag she could use to sleep on the floor. Instead she was told to climb into my window box and snuggle up to me. "Nicholas won't mind," someone had said.

Some time later I was dreaming of having a great deal of pleasure. I slowly woke up to see a stunningly beautiful young cis-woman with her mouth around my penis. The light was dim and her features were indistinct. I pushed the curtains slightly aside letting light from the street in. Thinking it was a dream I gazed at her without surprise, but said, "Who are you?" She answered, "What does it matter?" and sat up, striding my body with her hips she pushed my penis into her vagina and started to moan so sweetly, that another woman who had been with her boyfriend, sleeping in the loft some 15 yards away said, "Don't wake him up, he really likes to sleep!" Now awake, I saw her glistening and perfectly round breasts jiggling before me. As I caressed them I came to the full realization that it was not a dream and that I was being fucked by a most beautiful strawberry blonde. She slowly shook and slid to my side. I then cautiously moved over her and entered her. After that we fell asleep. This was a typical "adventure" in the Cockette world.

Some hours later I awoke and in the afternoon light gazed at her. I got up and made coffee as no one else had come to from the night's stupor. Going back to my window box with two cups, I found her now awake. "Hi," she said. "Tea please." After we had something to drink, we began caressing and this led to another bout of sex. After this we got up and went to the kitchen area that was open to the studio, that is, there was no wall separating the kitchen stove, shelves, refrigerator, etc. from the studio. There was a table that covered the trash bin on hinges that was used as a preparation area and temporary table with stools, and in one corner a table and chairs against one external wall with a doorway to the bathroom and toilet. Opposite this on the other external wall was a doorway to the back rooms and over it a stairway to the upper rooms. I think Scrumbly and some others had done the design and work, but I am not sure.

Karen (the name of the bed visitor) and I prepared a breakfast of eggs and toast and I introduced her to those who had not met her the night before. Karen filled in some background, telling me she was from Georgia. I remarked that my girlfriend at the time was from Georgia and Karen asked her name. "Melanie Mathias," I said and Karen looked a bit uncomfortable. "I know Melanie," she said slowly, "In fact, she was supposed to meet me today." I was now more than curious, "Like, you are friends with Mel?" "Good friends, we grew up together." "She is going to kill me!" I murmured. But Karen smiled and reassured me that Melanie would be fine with things. She was never possessive, she said of Mel.

Later Karen called Melanie and arranged for her to come to Valencia street. Now I was uncomfortable. An hour or so later Melanie arrived, she flew in the front door and turned directly to my space, she was familiar with it as she had been there many times before. Karen and I were sitting in the window box and Melanie shrieked, "You fucking slut!" in our direction. It was obvious that Karen had over-estimated Mel's tolerance for sharing lovers. Then to Karen she said, "You always do this, my boyfriends are fair game." To me she said, "Asshole, why did you fuck my (formerly) best friend?" I explained that I was asleep when it started and I did not know she was Melanie's best friend. After some discussion, where it was clear that neither Karen nor I knew the connection before the sex, Melanie calmed down and began an interrogation us about our future intent. I tried to be non-committal to future possibilities. It was clear to me that Karen was much more attractive physically than Mel, but Mel possessed such a dramatic and beautiful personality that there was no comparison. Karen just said, "If he is yours, then that's it." Melanie, then looked at me, and I said, "Mel, we are lovers, but not "old marrieds," If you want I will not do it again with Karen." "That's what I want, but I cannot trust you two, I can just look at you and know this is not over." It was over, as I saw Karen a couple of times after but she left town and I only saw her once after that several years later when she showed up at the Goodman Building suffering from goiter and in San Francisco for treatment.

I relate this event as an example of how people's feelings about love, sex and relationships were not uniform even among gay, straight or bisexuals. Things were quite fluid, and relationships soured, grew and died as with most people no matter what people's philosophical intent. The same situation had existed even in the hippie and political communes in the 1960s and early 70s (See my ethnography, Caldararo, 2019 for detail). Unfortunately, you had the mastermind controllers like Manson who try and control sex and become alpha male female owners, as well as the forced communalism of some groups where I recall people who were in love pressured to have sex with others to break their "sex property." The description of Manson by Timothy Leary (1976) is a good one on this where Manson's theory of fear becomes a behavioral doctrine of domination.

Cockette meetings to discuss a new play or finances could become quite funny, tragic or absurd. In the middle of one, Goldie Glitters, a main member, had an epileptic attack. The table around which some 10 or 15 people were seated and covered with food, drink, books and papers was overturned and several people ended up on the floor struggling to help Goldie. Once the fit was over, and Goldie had recovered, the meeting continued. Like many of the Haight-Ashbury communes where food was acquired from cheap outlets like Dented Cans, or from the Food Conspiracy where communes combined cash every week to buy food in bulk from wholesalers or farmers, the collective life of many of the Cockettes was ordered by agreed upon schedules and work details. I say, "many of the Cockettes" because one of the great distortions of the Weissman/Weber movie was to depict all the Cockettes as drug addled welfare cheats. This was simply untrue, many like Inez had full time jobs and while some were on SSI, and lived together cheaply sharing rent and food, others had no jobs or welfare, were wealthy or did what they could to get by.

Weismann contacted me in the late 1980s or early 1990s about film of Cockette performances he had heard I had made. He came to my home and we had a long talk about the film he wanted to create. I had worked for the Cockettes as a stage hand, painter, script writer and editor (for Martin Worman, especially) and in the process had taken some 8mm film of performances from back stage. Other film existed, especially that by the promoters of the Cockette debut in New York, but most of these were lost, inaccessible for useless according to Weissman. He was disappointed as some of the reels of film I had taken before going to the New York debut were stolen at the Chelsea Hotel where the Cockettes were lodged. Everyone had given their valuables to the Hotel clerk to place in a safe when we checked it as we were told it was the only prudent thing to do. But everything was stolen immediately and the Hotel manager told the Cockette manager, Sebastian, that the clerk on duty had denied taking items and the slips of paper we were given could have been written by anyone.

One of the common distortions of who the Cockettes were is the impression that they were all gay, or all men. While Pam Tent (the Cockette, Sweet Pam) Fayette Hauser and others have produced lists in their books on this, my own experience is close to these but I kept no lists myself. It is generally recognized that the Cockettes were multi-racial, usually focus is on Sylvester (one of my great friends in the group and a tragic loss in his death) and Reggie (also a great friend), and Linden (a brilliant performer) but there were many others. Link looked so ambiguous that he could pass for male, female, Asian, Mexican and Native American. He claimed to be a full-blooded Seminole. Sometime later in the mid or late 1970s I heard he had gone to Laos to find some aboriginal musical instruments and never returned.

But "ambiguous" applied in other senses as well. Some members of the Cockettes were ambiguous in appearance and behavior. One, named Markel, looked much like a woman but always dressed as a man. Everyone responded to he/she as if she/he was a woman and her voice was that of a female. Markel performed usually as a woman and had what appeared to me to be a female body. I only saw her naked once at the Chelsea Hotel where we were limited in bathroom space given that we had, as I recall, 2 rooms for some 20 people. Markel had come into the bathroom while I was drying off from a shower. She disrobed in seconds and jumped in the shower. The glimpse of her I had gave me, a trained physical anthropologist, little doubt of her biological gender. But again, performance was all, and she pulled off a persona of "no gender" quite well. When she had a role in a play as a woman, often in a dress of some elaborate type like a cocktail dress, she produced all the mannerisms typical of a role as female of the time described, as in the Cockette show, *Pearls of Shanghai.* As Goffman (1959) argues, the combination of role, manner and context create a front that people relate to as a common expectation.

In meetings Markel would generally ask questions and contributed in a rational fashion, but many people thought she was rather crazy. I was told by more than one person that she had been in a mental institution and had hallucinations of a bizarre kind. She once suggested an addition to a playbook where anarcho-syndicalists would aid aliens to save the characters in the narrative. She had listened to my discussion of the Civil War in Spain in the 1930s and guessed that I was sympathetic with the anarchists. But her topic of aliens was not unique, it recurred at various times and she seemed convinced that aliens watched human activity and were involved in rewards and outcomes. Someone once asked her what the aliens wanted, but her discussion was so complex I could not follow it.

"They are watching," was all I recall, but not in an entertainment or malicious way, more like guiding or pursuing, in the sense of how people might watch a plant grow, though that also is a rather distortion I think of her meaning. She usually avoided conversion, especially if it was argumentative, and shied away from confrontations.

In watching her behavior over the years, I came to think she was entirely sane. Like most women she was constantly being hit on or threatened by men and her response was to make herself plain and unobtrusive. When she dressed up for a play in the role of a woman she was often quite beautiful, and on stage she was confident and active. But once off stage she was obviously uncomfortable and if approached became incoherent or just silent. I was once stunned by how sexy she looked in one outfit, a tight-fitting gown for a scene that called for it. When the scene was over I complimented her, but was clearly interested in her sexually. She just said, "Nikki, you are a funny guy, not my type."

With my psychology background, and at this time I had had some six months of work at a state psychiatric hospital and had completed my B.A., I came to the conclusion that her odd behavior was just a ruse to prevent confrontations, and especially those with men. Otherwise, she might be really shy, but clearly very smart. I describe Markel in detail in this way to make the point that among the Cockettes many people's stories and personas were a work in progress as both performers and as characters in the strange and multi-layered reality that was the Haight-Ashbury and later Castro as drugs, fame and fortune progressed. Some years later I read Oliver Sacks semi-autobiographical work, *Hallucinations* (2012) and thought how he would certainly have fit in with the Cockettes.

I came into contact with the Cockettes late, as I used to go to movies at the Palace Theater in the early 1970s where mainly Chinese films were shown. I was initiated in this film genre by my girlfriend who was Asian American. Most of the films shown were fantastic legends of early China where people flew through the air and did remarkable feats. She thought they were funny, psychedelic and high camp. But one night we were leaving and a crowd was forming for a late show. This was a movie screening of what might today be called, art films, some foreign, some erotic and some cult films from early Hollywood and elsewhere. Since it was a hippie-like crowd with students we quickly found friends in line and were convinced to stay, but only after we got espresso and a croissant-like thing filled with something tasty and greasy.

## Membership, Who's Who and Hangers On

So later that night, after the "art movies," the Cockettes came on stage and did a rather bizarre, camp and DaDa rendition of some 30s movie or movies. I was hanging out waiting for my girlfriend who was in the toilet, and, one of them, Link came up to me. He had stripped himself of all the drag, glitter and makeup of the show and asked me for a drink of the Coke I was holding. We had a chat about the show and he invited me to the house he was living in at the time the next day. The next day, I came to their door and Tahara answered my knock. "Ha, the pretty boy in the lobby." She said. "Everyone noticed you." "Wow, Great, I guess." I said. She/he led to up the stairs so typical of San Francisco Victorians and up a hallway, then pointed out a doorway and said. "He's there." They had taken off the doors to most of the rooms as I recall, or it might have been only this one. I had seen this before in other communes. The room had been divided into sections both horizontally and vertically, so there was a main horizontal division made by a loft that cut the room in half. Given that the ceiling was about 12 or 15 feet, this left plenty of room for two levels. Then the room was divided in these into two sections so that there were 4 living areas. Link's was on the bottom left.

## Transsexuals, Transvestites, Cross-Dressers and Bathrooms as Sacred Spaces

I came into the room and someone, I think it was John Flowers who was both an actor, writer and set designer with Fayette, said Link had just left. John then said, "You're Nicholas, right? We met." One of my girlfriends at the time was bisexual and I had often gone with her to lesbian and gay bars, usually dressed as a woman. I, like Miss George, had often been dressed as a girl as a child. My sister was older and her cast offs came to my use. I was not confused about my sexuality or gender but rather mystified by the special nature of femininity. My sister and her friends found it amusing to play with me and we often went out together with me in female clothes. This background made my involvement with the Cockettes rather a natural. But I was often uncomfortable around some of the Cockettes and their friends. If Hibiscus could proclaim that acting and being were the essential element for the desired effect, the problem was in the effect. I was unsettled by the way many of the Cockettes portrayed women, it was more than a silly stereotype. There was usually a mocking and insulting nature to the act and it was something I had come to dislike about "drag queens" and "transvestites" in general at the time. I loved and respected women, spending a day with a woman friend in book stores or library was as rewarding as spending a night cooking or clubbing and morning lolling in bed with a girlfriend. Goldie Glitters once said of me that "you lack the milieu of a queen or homosexual." And it was true, I was neither a "queen" or a homosexual, I had no exaggerated gestures that Goldie felt were markers. Of course, she/he was wrong, many homosexuals then, and bisexual men, had no outward gestalt of gayness, if you will.

The other part of this is the other's perception. Acting is the intention of the performer, how the actor is seen is another worldview. When I was a child, probably about 6 or 7 I had gone with my sister and her friends to the town's bowling alley wearing a dress and my "Prince Valiant" wig I had got as part of a Halloween costume. It was a place where my sister and her friends liked to visit and sometimes bowl. After a certain amount of time I needed to use the toilet and approaching the men's bathroom, I was approached by an older woman (at that age one cannot discern really the age of anyone over 10 I think), who grabbed my arm and said, "No, sweetie, you have to use the women's toilet, it is here." She then led me to the women's toilet and smiled as I entered. I could not tell if she recognize me as a boy in a dress or not. However, from that day on whenever in a dress I just went to the women's bathroom.

Some years after, on Halloween, a girlfriend took me to a straight party one night before going to a club. I was in a dress as I often did with heels and full makeup. A man was there in a dress and obvious garish makeup and wig making a variety of comic acts stereotypic of how women are often mocked. When my girlfriend saw him she was not happy and, being a feminist, expressed her displeasure. Someone near by joked about it and the man came over. My girlfriend knew him and introduced us, when he realized I was a man, he became angry and started to threaten me. It appeared irrational and out of place given his dress, but I realized the problem was that I was not mocking women, but simply dressed in a serious fashion and he was probably sexually confused if not attracted and having a "homosexual panic," as it used to be called, questioning his masculinity. Absurd as that was even on Halloween. When we disengaged my girlfriend introduced me to other people and one woman said, "You look great as a woman, but you don't act as feminine as you should." My girl friend then said, 'Niccolo is just being himself, he is just in a dress. He is his normal self in one kind of outfit or another.'

Yet the aroma of the time was creating a number of other critical responses to cross-dressing. Many feminists saw cross-dressing and transvestites as caricatures of women, that men acting as women was like white people putting on blackface. A few years later a girlfriend took me to a meeting of local feminists who had come to know that she, a prominent feminist, was dating a man who often wore dresses but also had a connection with the Cockettes who they considered to be stereotyping women. I was reluctant to go, but was convinced it would be a friendly discussion. She was right, I was questioned sometimes heatedly but respectfully. I tried to make the point that I had little to do with drag queens or transvestites as most acted. I had no personality need of "acting out" being a female and I felt I could not speak for what other people who were biological males were thinking when they wore women's clothes. Nevertheless, given the political context and the intellectual orientation of the group, there seemed to be a need to construct some theoretical explanation for my behavior. Not all those present agreed, but the overall view was that I dressed as a woman to attract women, appeared not to be an oppressing male, yet once a woman was hooked they were sure I was as much a male supremacist as any other male. No objection by my girlfriend could dissuade them, she was simply the victim of the false consciousness my theoretical ruse had produced in her. In this group the proof that I had dominated her was that she was having sex with me. Since all sex with men was rape to them, she was therefore dominated. My dressing like a woman was just a pretense. Others who were not convinced, simply asserted that there were always women who dominated effeminate men, but could not reconcile that with the fact that my girlfriend had no history of that (she was married and had children), though agreed that no one could be sure how people acted in private. "Everything is political in the end." Said one.

This kind of sexual political analysis was also paralleled by other political debates. At this time in the evolution of the Cockettes, there was a fissure growing among the members, one that was apparent across the Haight-Ashbury and the entire "counter culture" at the time of the early 1970s. Hibiscus was heavily influenced by the idea of leveling and communal living. Also, he rejected authorship and stardom as an individual status. It seemed to me when I listened to what others said of him, that he was erratic and dogmatic in theory and practice. The only personal interaction I ever had with him was when I was discussing belly dancing with someone and he got so excited he wanted me to do one in a show he was writing. I was studying the dance, both from an academic perspective as its relation to women's expression, role in child birth, cultural differences (e.g., Turkish vs Egyptian) and the variety of clothing and music. I did perform my interpretation of a more Berber variety for one of his shows.

The "personal is political" view dipped into all aspects of life and the communes had effects beyond their number, note the considerable extent of the hippie movement and the Food Conspiracy-like enterprises, in various sectors of the American population. Because people in households participating in the Food Conspiracy had to send representatives to meetings and to the bagging and other tasks for the food distribution and financing, there were many opportunities for information and influence. Some communes were highly political and dogmatic, some Marxist, some anarchist and some like the commune, Cauliflower, political in terms of gay rights and expression.

It seems an exaggeration or confusion to say that a man entering a woman's restroom is a political act. Some have made this reference. Rothman (2016) describes how the territory and use of the woman's restroom, bathroom, powder room, etc. has become a major political issue in America linked to transsexual status and sports performance. The

American Psychological Association (Anon, 2023) has made a determination on even "all gender" restrooms. But again we are faced with the same problem, what is a woman? Certainly many androgen sensitivity XY people are "women" in almost every impression, and since the god of those who most rail about the bathroom issue has not defined the nature of woman genetically or who should be allowed in what (since I am a scholar of religious texts I can speak on this see for example: Caldararo, 1994; Caldararo, 1994a; Kahle and Caldararo, 1986) we have a dilemma of opinions. I have gone many places with my wife (a cis-woman) where we are faced with bathrooms for men and women where there is a huge line for the woman's bathroom, but hardly any for the men's. She has often used the Men's bathroom while I stood guard preventing others from entering. Most men in this situation found it amusing and there has never been a complaint. There are several reasons for this, one is that in the men's bathroom it has been traditional to provide a "urinal" where a man can urinate standing up. Whether this reduces the tendency of men to urinate all over the seat of a toilet is questionable.

Nevertheless, in women's bathrooms there are only toilets. This requires more time to disrobe to allow for use of the toilet. Also, men seldom clean themselves after urinating in either a urinal or toilet. Thus the time involved for men or women's use of a restroom differs substantially. This brings up the point of the social function of the restroom or public bathroom. Many people, especially young women use such facilities as a platform for socializing. But such social activities have a long history as in the public baths of the Romans and Ottomans, and the restroom of the 19th century (Hoagland, 2018). It can be argued that the restroom is both a socially inclusive place and one of exclusion, where certain types of bodily display and language take place in the limnality (semi-sacred) context. While social divisions (class and race) have limited use to certain individuals and times of day (most public restrooms are locked at night); exclusion and definition of the right to use a bathroom has a number of socially revealing aspects, not only who is excluded, but why.

This does relate to custom and learning as some have complained. I was with a group of feminist women one night at an outdoor even in the

1970s. The promoters who put on the event had not provided sufficient rental toilets for the crowds of people attending. We were milling about like hundreds of others and one of the women pointed out that many men were simply urinating in public. She said, “Woman can’t do that.” I said, “Why not, just do it. Who cares.” Another woman said, “women cannot urinate like that, it just dribbles out and makes a mess.” I said, “Women have been urinating without toilets for thousands of years.” Another said, “Yes, but if you are naked or wearing a skirt you can squat and do it ok. The urine comes out in a strong stream and as I girl I used to try and aim it.” None of the women present was wearing a dress. I conjectured, “What happens if you pull back the labia and lean back and press on your bladder, that would make some difference.” None of them had ever tried that, so two dropped their pants, pulled back their labia and tried pushing the urine out. The result was varied by the degree of hubris in my mind. One woman was more successful in holding her labia back and stretching the skin around the urethra. The other was less so. The former succeeded in producing a stream of urine that was like a man, the latter dribbled and spraying making, “a mess,” as she said. It is interesting that a 2011 study by John Hopkins found 35% of men sat down on a toilet to urinate. A 1981 study by Charles Gerba found that men and women visited restrooms about the same number of times a day but that women spent about twice as much time as men in the rooms.

A study by Formica (n.d.) found quite contrasting attitudes about bathrooms. While men and women both considered most public bathrooms dirty (72% of women, and 64% men), females desired more time to use facilities, to undress, wash and dress. 40% of women spent 1-2 minutes in a bathroom, 22% 3-4 minutes. Men were in and out with as little time possible with 41% of men were in an out in about 30 seconds. Surprisingly, the difference in hand washing between men and women is small, 35% of men and 29% of women report a lack of hand washing. However, a Michigan State University study in 2013 found that 61% of women and only 40% of men washed their hands after using the toilet, but if there was another person in the bathroom the percentage of hand washing went up significantly to 90% for women and 75% for men (Daly, 2010).

## Sex as a Means of Communication and Celebrity and the Lack of It

Cockette parties were of two kinds in my experience. There were the after show parties that were attended by friends and mostly performers and staff. These began late, usually after 1 a.m. in the morning and could go on into daylight. They often relocated in restaurants for coffee and food. While there was some discussion of the performances that had taken place, they were usually oriented for alcohol, drugs, sex and just plain fun. I cannot recall any sex that I had that was not involved in discussions about the plays after though.

While there was plenty of mindless sex going on, especially among members away from the performances and parties, people were generally open to sex at any time, place or with whoever. But as I said above this varied considerably. Many people were brought into the enterprise through sex, people were met at bars, on the street, in restaurants, wherever, and would show up at meetings and, sometimes, on stage. There were no formal auditions as far as I could tell, though once I was asked to come to a Cockette house and when I got there two of the general organizers, including the manager, Sebastian asked me to sing, dance and provide an example of my talent. Shocked and surprised, I remarked that I was not a performer. Therefore, one might say that there were ad hoc auditions.

The second kind of Cockette party was individually centered on the housing situation. As not all the Cockettes lived together there were several types of living arrangement and these presented different opportunities for parties. Often I would say a dinner could become a party and a rather small group of 8 or 10 people would eat, dance, sing and drink or take drugs together. Sometimes these parties would begin as a work meeting and then transition into a party. As I have said, the purpose of life was to thrive and have fun.

## References

Anon, (2023) "Universal design all-gender restrooms," *American Psychological Association*, https://www.apa.org/topics/lgbtq/all-gender-restrooms.

Anonymous, (n.d.) "Washrooms: gender differences," https://www.formica.com/en-gb/articles/commercial-interiors/washrooms-gender-differences.

Caldararo, N. (2022) *A Mental Ethnography,: Conclusions From Research in LSD,* Cham, Switzerland, Palgrave/ Macmillan.

Caldararo, N. (2019) *An Ethnography of the Goodman Building, the Longest Rent Strike,* Cham, Switzerland, Palgrave/Macmillan.

Caldararo, N.L., "Storage conditions and physical treatments relating to the dating of the Dead Sea Scrolls," *Radiocarbon*, v. 37, n. 1, 1994:21-32).

Caldararo, N., 1994a, "The lost libraries of the Americas," *Library History*, v. 10, pp. 9-18.

Crumrine, N. Ross and Marjorie Halpin (1983) *The Power of Symbols: Masks and Masquerade in the Americas,* Vancourver, University of British Columbia Press.

Daly, Judy (2010) "Public Handwashing," American Society for Microbiology, https://www.sciencedaily.com/releases/2010/09/100913111121.htm.

Fraley, M. (2016) "Gwen Araujo murder 14 years later: Transgender teen's killers face parole," *East Bay Times,* Oct 14.

Goffman, Erving (1959) *The Presentation of Self in Everyday Life,* New York, Penguin Books.

Hauser, Fayette, (2020) *The Cocettes: Acid Drag and Sexual Anarchy, 1969-1979,* Process Books.

Herdt, Gilbert, (1987) "Selfhood and discourse in Sambia dream sharing," in Barbara Tedlock (ed.) *Dreaming: Anthropological and Psychological Interpretations,* Cambridge, Cambridge University Press, pp. 55-85.

Hoagland, Alison K. (2018*) The Bathroom: A Social History of Cleanliness and the Body*, Santa Barbara, Greenwood.

Kahle, T.B. & Caldararo, N.L., "State of preservation of the Dead Sea Scrolls," Nature, 8 May 1986, v. 321, n. 6066:121-2.

Leary, Timothy (1976) "Meeting Manson," *Oui Magazine*, August, 47-102.

Lewis, Gilbert (1977) "Fear of sorcery and the problem of death by suggestion," in *The Anthropology of the Body,* John Blacking, (ed) London, Academic Press, pp. 111-144.

Magee, Carol (2012) *Africa in the American Imagination.*

Marston, Wendy (1999) "Scientist at work: Charles Gerba; on germ patrol, at the kitchen sink," *The New York Times,* Feb. 23. https://www.nytimes.com/1999/02/23/health/scientist-at-work-charles-gerba-on-germ-patrol-at-the-kitchen-sink.html.

Randi, James (1993) *Conjuring*, St. Martin's Griffin.

Rivers, W.H.R. (1920) *Instinct and the Unconscious*, Cambridge, At the University Press.

Rothman, Joshua (2016) "The politics of bathrooms," *The New Yorker,* May 14.

Sacks, Oliver (2012) *Hallucinations,* New York, Vintage Books.

Tent, Pam, (2004) *Midnight at the Palace: My Life as a Fabulous Cockette,* Alyson Books.

Travers, Mark (2024) "3 Motivations behind 'cosplaying,' according to a psychologist," *Forbes*, March 25, https://www.forbes.com/sites/traversmark/2024/03/25/3-motivations-behind-cosplaying-according-to-a-psychologist/.

Trivers, Robert (1985) *Social Evolution,* Menlo Parl. The Benjamin/Cummings Publishing Co.

# Chapter Four

## Why most Birds lack penises and what it means for being male and domination for most Vertebrates

While there has been considerable debate and theory production about why humans have such big brains (Caldararo, 2017), one simple answer some people produce in jest is that the big brain makes for rapid ecosystem destruction. Perhaps the human explosion of power has been so fast and so thorough that extinction may be our finest product. In bird species, whose males lack a penis, there is a necessity of cooperation to achieve coitus and successful impregnation. Obviously this is imperative for species survival and if the rewards of mutual sexual achievement provide sexual advantage, either for reproductive success or group solidarity (as seems probable in coupling in birds and their examples of social grouping) then the development and evolution of conceptual sexual rewards and identity in partnership would be a clear avenue. If this were true one would imagine that today humans would be doing a better job of satisfying each other. This applies not only to humans, though as Eberhard (1985) has described in great detail (and Darwin noted over 100 years ago), the male genitalia is one of the most complex organs of an animal's body, yet its purpose is one of mechanical simplicity. This idea of the partnership of gender is a central argument of McKenna (1992) associated with his emphasis on the use of psychedelic drugs in pre-modern societies to instill more plastic social relations and complex consciousness. The idea of sustainable social organization and ecological adaptations are not new and do not depend entirely on gender equality or the use of psychedelics as I have noted in my review of human economic adaptations (Caldararo, 2014) and is found also in Herskovits (1952), though Leacock and Nash

(1977) make a strong argument for this idea.

There is evidence to consider that some human societies were once more balanced in gender relations and power. This might have been true before the invention of urbanism and states, of perhaps 5,500 B.C.E. but according to Jaspers (1953) the spread of ideas was significant, though his argument is more convincing as the invention of ideology and its role in mass organization. A date around 2,500 BCE if we can use Malinowski's data on the Trobriand Islanders (Malinowski, 1929) as a point of reference compared to the world denying religions that became dominant after Jasper's Axial Age. However, it is always dangerous to attempt simple, single factor explanations for any evolutionary or historical event (Boas, 1920; Simpson, 1949).

Darwin's idea of female selection is one idea applied to differences in physical appearance or morphology in animals. Often referred to as sexual dimorphism, it has been used to explain differences in size, color, and a variety of other differences or secondary sexual characteristics like human breasts. Though both males and females have structures that appear similar, e.g., nipples and can express milk, they usually vary in size between the sexes (though many females and males overlap in size as in other characters). If we consider the penis one of these we can note its variation in size and shape (Burton made a study of this in the 19th century, see Kennedy, 2009), but the most immediate aspect of the penis is the use of it to change the female, especially in most vertebrates by injecting sperm into them. In some rodents, mainly mice and voles, an impregnated female can nullify the injected sperm, known as the Bruce Effect (Schwagmeyer, 1980).

## What is its Origin and Why You can do Without it?

The penis has an ancient heritage and yet it has disappeared from some vertebrate lineages (Brennan, 2016). It was thought that all terrestrial amniotes had penises with the exception of the Tuatara, a lizard (Sphenodon punctatus), but recently it was noted that their embryos have genital swellings that indicate a uniform ancestry. The tuatara male lacks an intromittent penis (that is an external organ to deliver sperm). Definition of male and female often revolve around the possession of one gender of a penis. This is complicated in some cases as in a cave insect (Neotrosusgla) where the females possess a "female penis" or intromittent organ for coitus (Yoshizawa, et al., 2014).

Of the nearly 10,000 species of birds only about 3% have penises (Yong, 2013). We have to wonder why, what is the selective advantage for a penis and for not having one, or, that is, in an evolutionary context, losing them? A pleiotropic theory has been proposed by Mayr (1963) and was modified to clear some objections by Arnold (1973). Here the variations between penis and vaginal structures and secretions results not directly from selection but from other factors where effects on the reproductive mechanisms are a secondary consequence. This blends into conflict of interest between the male and female in a number of factors like energy expended to produce structures of chemicals, interference with survival as in big penis and running or exposure to injury and infection in a mechanical sense (Trivers, 1972).

The loss of the penis in male birds has been addressed by a number of theories, for example, the reduction in flight drag or the nature of sexually transmitted disease in birds due to the liability of bird species having a cloaca that is common to both the genital and gastrointestinal systems. Both appear to be weak as Briskie and Montgomerie (1997) argue. A female choice hypothesis is presented by them where males without a penis allow for greater female control over investment in offspring. Yet the variation in Bat penis structure, in the bacula (Hamilton, 1949) is remarkable and being that they engage in flight this should be an effective argument against the loss of the penis in birds being due to flight restrictions.

Another theory is the "Genitalic Recognition" theory where the penis has evolved in the male to stimulate the female and thus maintain a species identity interaction. This is also related to a broader hypothesis of the "lock and key" theory that variations in penis and vagina evolution drive speciation and success in mating (e.g. Hammond, 1981). Most birds then, must bring their genital areas together in what bird behaviorists call, the "cloacal kiss." However, a most interesting fact of genitalia conformation is that in many insect classes the female genitalia is nearly uniform across groups of species while the male is modified and often elaborated (Eberhard, 1985).

It is obvious that the lack of penetration by the male penis likely reduces some forms of infection, but the sperm and related fluids and lubrication would still seem to be a source of infection and parasite transfer. An other clue might arise from the fact that birds and penises can generally be divided into two groups, Galliformes, land forms without a penis and Anseriformes, waterfowl with them, given also that many aquatic animals also have penises. Some mammals have a bone in the penis as mentioned, but others a coat of spines. Snakes, lizards and kangaroos have two-pronged penises, while echidnas have four-pronged ones. Mammals inflate their penises with blood, birds use lymph and alligators have a permanently erect penis (Yong, 2013; Eberhard, 1985). The variety of penis form is quite substantial. The idea of female choice and the loss of the penis in birds has some support among ornithologists. It would reduce male ability to rape as is present in ducks, also it gives females more control of mating as they must cooperate to produce coitus (Herrera, et al., 2013). This gives an explanation for the relatively common monogamy in birds, while not perfect it is certainly more secure than in humans.

So if we cannot rely on the possession of a penis to define maleness or femaleness, what about sex chromosomes? Most mammals have these and possession of two X chromosomes and one X and one Y chromosome usually indicate female and male gender. There are significant variations in number of chromosomes and gender status.

**Mammals**
XX female XY male
XY with
XO XXX female XYYY Male
XXY female appearing Klinefelter's Syndrome – hermaphrodite
**Platypus**
**5 X and 5 Y**
**Birds (General)**
ZW female ZZ male
**Some Insects** (e.g., Drosophila)
XXY female XO male

**Sources: Moody, 1967; Veyrunes, et al., 2008)**
**Table 1 Some Variations in Chromosome Type and Gender**

We might argue that the penis is not to blame for climate change, but rather the lack of restraint derived from child-rearing that leads not only to aggressive, and in many cases, destructive behaviors as well as spousal abuse. Margaret Mead (1928) attempted to create a paradigm to understand how child-rearing techniques molded behavior and resulted in specific types of social behavior. To change the trajectory of human society and ecological disaster will, I think take more than changes in child-rearing. Especially since there is no consensus on the nature of climate change with some ideologies holding beliefs diametrically opposed to scientific evidence. No matter what the threat, e.g., temperatures too high for human survival (Vecellio, et al., 2023) or substantial changes in pathogens (Mora, et al., 2022), these ideological explanations for disbelief will continue and some will lean on futurists who predict greater human innovation and adaptability to counter climate change.

While we might characterize this attitude as a healthy approach to solving a threatening problem, it can also be seen as rooted in the psychology of denial that is structured in child-rearing where people are enveloped in a system of conditioned responses to contradictions of their training as their cultural cosmology was learned (Kardiner, 1963). In 21st century, post-industrial society with an internet and electronic devices that can determine identity in complex fashion as, perhaps never before in a context of anonymity, having a penis seems less important than what a brain can imagine. Advantage and conditioning based on cultural precepts will continue to dominate the world order unless some major shift in thinking can occur, perhaps one as great or greater than the Bubonic Plague. The creation of a sustainable culture and technology will be the crisis point to evade technological disaster as in not only climate change but nuclear, biological war and related accidents. The internet is filled with questions from women about whether they should submit to their husbands and how much volition they should surrender. Answers are also there in the 10s of thousands of advice channels and these are generally religious in nature. Whether a patriarchal or matriarchal society, democratic or authoritarian is best for sustainability will be addressed in the next chapter.

## References

Briskie, James V. and Robert Montgomerie, (1997) "Sexual selection and the intromittent organ of birds," *J. of Avian Biology,* v. 28, n. 1, March, 73-86.

Brennan, Patricia L.R., (2016) "Evolution: One Penis After All," *Current Biology,* v. 26, n. 1, R29-R31.

Bogin, Barry (1999) *The Growth of Humanity*, New York, John Wiley & Sons.

Cakdararo, Niccolo, (2014) *The Anthropology of Complex Economic Systems: Inequality, Stability and Cyvles of Crisis,* Lanham, Lexington books.

Hamilton, W.J. (1949) "The bacula of some North American vespertilionid bats," *Journal of Mammalogy,* v. 30, n. 2, May, pp. 97-102.

Hammond, P.M. (1981) "the origin and development of reproductive barriers," in *The Evolving Biosphere*, eds. P.H. Greenwood and P. L. Forey, Cambridge, Cambridge University Press, 37-48.

Herskovits, M.J. (1952) Economic Anthropology: A Study in Comparative Economics, New York, Alfred A. Knopf.

Herrera, Ana M. Simon G. Shuster, Claire L. Perriton and Martin J. Cohn (2013) "Developmental basis of phallus reduction during bird evolution," *Current Biology*, V. 23, n. 12, June, 1065-1074.

Jasper, Karl (1953) *The Origin and Goal of History*, London, Routledge.

Kardiner, Abram (1963) *The Psychological Frontiers of Society*, New York, Columbia University Press.

Kennedy, D. (2009). *The Highly Civilized Man: Richard Burton and the Victorian World*, Cambridge, Massachusetts; London, England: Harvard University Press.

Leacock, Eleanor and June Nash, (1977) "Ideologies of sex: archetypes and stereotypes," *Annals of The New York Academy of Sciences*, v. 285, 618-645.

Mead, Margaret (1928) *Coming of Age in Somoa,* New York, William Morrow & Co.

Moody, Paul Amos (1967) *Genetics of Man*, New York, W.W. Norton.

Mora, Camilo, Tristan McKenzie, Isabella M. Gaw, Jacqueline M. Dean, Hannah von Hammerstein, et al., (2022) "Over half of known human pathogenic diseases can be aggravated by climate change," *Nature Climate Change*, v. 12, 869-875.

Schwagmeyer, P.L. (1980) "The Bruce Effect: an evaluation of male/female advantages," *American Naturalist*, v. 114, 932-938.

Vecellio, Daniel J., Oinqin Kong, W. Larry Kenney and Matthew Huber, (2023) "Greatly Enhanced risk to humans as a consequence of empirically determined lower moist heat stress tolerance," *PNAS*, v. 120, n. 42, https://doi.org/10.1073/pnas.2305427120.

Veyrunes, Frederic, Paul D. Waters, et al., "Bird-like chromosomes of platypus imply recent origin of mammal sex chromosomes," *Genome Research*, July, v. 18, n 6, 965-973.

Yong, Ed (2013) "How chickens lost their penises (and Ducks kept theirs)" *Smithsonian Institution Magazine,* June 6, https://www.nationalgeographic.com/science/article/how-chickens-lost-their-penises-ducks-kept-theirs?loggedin=true&rnd=1687905065659.

Yoshizawa, Kazunoiri, Rodrigo L. Ferreira, Yoshitaka Kamimura and Charles Lienhard, (2014) "Female penis, male vagina, and their correlated evolution in a cave insect," *Current Biology*, v. 24, 1005-1010.

# Chapter Five

# Marriage variation and matriarchies as future realms

This chapter is not a comprehensive analysis or history of marriage in humans. Often discussions of marriage types begin with a variety of history. Often something like this: "in the dim mists of the past... there was no law just brute force and submission to it...the primal ancient horde knew no marriage rules but was dominated by lust and promiscuity..." Such fantasies were replaced in the 20$^{th}$ century with studies of primates, not that reference to animal societies had not been done before. In antiquity several authors had done so. The same can be said of scripture, where we find arguments about the meaning of sayings like, "go forth and multiply." This passage, while not from God (Genesis 1:28) comes from a man, Isaac.

## Ideologies of Marriage

To use the term, "ideology" in a work of anthropology, brings to mind Geertz's (1973) extensive discussion of the term. A belief in how things work on an absolute scale and a faith in the authority of the belief, is I think an adequate definition. This then covers those ideologies of religion, modern philosophies and of political philosophies. Whether the authority is Christ, Buddha, Mohammed, Marx or Hitler, the belief is established as a certainty. The same can be said of the ideas of life into which the ideas and faith of pre-modern people of the Bronze Age and before can be assumed by reference to those people who have given record of their non-industrial, traditional ways of life, for example, the Andaman Islanders (Radcliffe-Brown, 1964).

We have considerable archaeological and ethnohistorical evidence that prior to the invention of agriculture human groups grew in population slowly if at all. While there is considerable debate over whether population control was a conscious adaptation to the carrying capacity of the environment, the explosion of population after the Neolithic is unmistakable (Cohen, 1977).

There are a number of religious influences to marriage as well as economic and military. As regards status in a community, who one marries can be very significant not just for the bride and groom, but for the status of the children. Usually we see children possess the status of parents, modified by whether the family is in a matrifocal for patrifocal society. Your relatives and duties as well as status are determined by the lineage of the parents. In some cases this information is secret, as where certain aspects of power are relegated by birth, for example, among the Lovedu of South Africa (Vansina, 1985). Here the paternity of heirs was secret kept by elements of the royal office as the throne was passed down through the female lines.

The influence of religion is complex and even in so well known a religion as Christianity, its attitude towards marriage has changed drastically in the past 2000 years. The policy of early Christians towards marriage, I must say, related to the 'specific end of times' ideology current and thus among the Essenes (Joseph, 2018; Josephus, 66-67 C.E.; see also Jonas, 1958 and Du Pin, 1692). Men and women lived apart and sex as well as children were forbidden. One has to remember that early Christians were a diverse group, the transition from Jewish sect to separate religious entity took time and was partly also the history of the rebellion against Roman rule and the resulting partial diaspora of the Jewish population after the conquest by Titus (Sand, 2009). You can refer to the original history of this time from the hand of the Israeli general Josephus who was captured by the Romans and later wrote both a history of the Jewish War and a memoir of the time's political and religious issues.

The history of marriage among the early Christians is clouded in the lack of documents on this issue. Discovering the history of the role of marriage is in du Pin's text. The book (a different number of volumes are produced in the different editions both in French, Latin and English that I have seen. Mine is the original English translation of 1692 a text used by a family to record births and deaths from that date. It does have an index, but my reference to du Pin's work was not to provide a final authority on marriage, but to show clearly how the Church has reformed scripture over time. The book contains complete texts of documents believed at various times in the past 2000 years to be authentic writings or recorded sayings of Christ and/or the Apostles. The Church fathers, in concert (what were often called "councils") and by virtue of special study or decisions by particular Popes, regarded and often expelled from accepted Scripture letters, passages and sections of works attributed to Christ or his early followers. Today we call these texts the Apocrypha and different texts are considered such by different sects of Christianity and different scholars. For example, the Vulgate copy from the King James version. These were copies of volumes or writings, sayings, etc. that circulated among early Christian communities before printing. At times they were considered authoritative, then heretical or fakes. Differences of opinion by Church scholars are included in du Pin with full quotations of their arguments. It was originally published as a reference during the Catholic Counter-Reformation to answer charges by Protestants that the Church lacked a consistent body of scripture but had arbitrarily modified dogma to fit its needs (see the writings of Martin Luther or Erasmus).

Further questions can be found in McKnight, et al., *The Synoptic Gospels* or in Hans Jonas', *The Gnostic Religion.* The central point of discussion was how the production of children would hinder the return of Christ by dimming the "light" of the faithful and give power to the demiurge by the magnification of error. This debate is recorded in du Pin under the topics of the followers of Marcion and the Valentinians. On the other hand du Pin relates the condition of early Christians as living in a state of celibacy (page 202 of volume I). Since I do not have any other copies I note that it comes in his section on the "Abridgment of the Discipline" where he acknowledges records of the early Church in the first 3 centuries.

It should be noted that even in du Pin's time scholars spent a lifetime on a small section of documents to discover the meaning and intent of the scattered segments of the Church as it developed. This is enlightening as to the means by which people attempted to adapt a faith to changing conditions. While there is no lack of studies of the variety of human marriage or sexual arrangements (concubinage, prostitution, etc.) aspects of human sexual response may hold the key to stable and adaptable human societies of the future. We may find that the satisfaction of sexual desire, in both actuality, forms of worship as in recognition of beauty and pornography as well as plasticity (e.g., polygamous association, see e.g., Schippers, 2017).

Our central problem is the small amount of information we have about human bonding, whether for sexual purposes or reproduction. In the ethnographic record where it is less distorted by contact with the mass religions like the Judeo-Christian-Islamic and Buddhist modes, we have some interesting variations (Schneider & Gough, 1961; Radcliff-Brown & Forde, 1950; Goode, 1971). These are usually described and analyzed by anthropologists in the context of social structure and family/kinship relations as in polygyny, polyandry, gender avoidance and non-spousal sex, same gender marriage and sex and forms of adoption (Stone, 2006). As Leacock & Nash (1977) have argued even these shreds of human variation give us a remarkable sense of potential for a future of tremendous variation. The possibilities are there, can the human superorganism shift in time to make a difference. And it is not just in looking back tom models of what has been, recent sexual and family experiments from hippie communes to kibitz and Mormon super families of neo-polygyny show new trends in adaptability.

## References

Du Pin, Lewis Ellies (1692) *A New History of Ecclesiastical Writers...*, 1st English trans., London, Abril Staglle and Tim Shilbe.

Goody, Jack (1971) *Kinship*, Harmondsworth, Penguin.

Joseph, Simon J. (2018) *Jesus, the Essenes, and Christianity*, Waco, Baylor University Press.

Jonas, Hans (1958) *The Gnostic Religion*, 2nd Ed., Boston, Beacon Press.

Josephus, (66-67) *The Jewish War*, Trans. By G.A. Williamson, revised by E. Mary Smallwood, London, Penguin Books.

Mcknight, Scott, John K. Riches and William Telfordand Christopher M. Tuckett (2001) *The Synoptic Gospels,* Sheffield, Sheffield Academic Press.

Radcliffe-Brown, A.R. & Daryll Forde (1950) *African Systems of Kinship and Marriage,* London, Oxford University Press.

Radcliffe-Brown, A.R (1922) *The Andaman Islanders,* New York, The Free Press of Glencoe, 1964 edition.

Sand, Shlomo (2009) *The Invention Of the Jewish People,* London, Verso Books.

Schippers, Mimi (2017). *Beyond Monogamy: Polyamory and the Future of Polyqueer Sexualities*. New York, NYU Press.

Schneider, David & Kathleen Gough (1961) Matrilineal Kinship, Berkeley, University of California Press.

Stone, Linda (2006) *Kinshiop and Gender,* Boulder, Co. WesrView.

# Chapter Six

## Beyond Zero Population: Ehnohistory, Archaeology and the Khmer, Climate Change & the Collapse of Civilizations

A number of publications, both books and articles, have appeared in recent years attempting to prove that there is a correlation between global climate change and the collapse of complex human societies. Buckley, et al. (2010) is one example where emphasis is placed on climate over human activity. This chapter addresses a possible exception this emphasis of global effects. While societies under stress from weather changes may undergo significant economic and political change in response, Joseph Tainter (1988) has shown that a local systems analysis is best applied to determine the contribution of the internal dynamics of a society, rather than assign climate the ultimate effect. In this chapter ethnohistorical and archaeological sources are reviewed indicating varied responses to climate challenge. We find that some, like Khmer society, create initially a sustainable responses to local conditions and an exception to the proposed global collapse theory. However, failure of surrounding populations created pressures on Khmer society that destabilized adaptations to environmental stress forcing collapse.

Like the Bronze Age collapse in the Mediterranean (Braudel, 2001), local accommodation was insufficient for continuation on existing patterns. Others like the Japanese modified population densities and social forms to weather stress. Implications for today concern continued population growth worldwide and rising consumption are continued stresses that must be considered in the context of climate change. Pre-Neolithic population densities and modern technology could be a means of adapting human society to the constraints of quality of life and diminishing returns of technology. Loss of technology is also a major consideration, and while some climate scientists expect major reductions in energy production hampering basic machine culture, we must also expect loss of skills necessary to maintain mass industrial society. We have just lived through the Covid pandemic which caused not only a considerable disruption in work, but also in education and the development of social skills in children. We should recall that with the major collapse in antiquity, the Greeks forgot how to write and read their own script.

## Introduction: Climate Change Factors

The debate over contemporary factors and effects of climate change has created a search for evidence of climate change in the past and the effect of climate change on past civilizations. What can archaeology tell us about the end of human industrial society, and perhaps also human life? Will there be an Earth without humans, the end of life on Earth or a sustainable Earth with humans? Decades ago Rene Dubos (1973) pointed out that humans were capable of creating productive and continuously useful environments out of "wilderness" but this is a valued consideration. "Wilderness" is that area that is productive for those plants and animals found within it, but where humans have not established themselves. Dubos and others have noted that human arrival changes ecologies dramatically making them "human" (Dubos, 1974). He also argued that to survive humans must adopt a social structure that promotes a "steady state" of population and resource need (Dubos, 1970). I discuss this idea in detail as well as the contributions to it by economists like Daly and Boulding in my 2014 book on economics.

Most studies like that by Tainter (1988) focus on human detrimental effects on the environment, on climate change, pathogens or human inability to create resilient institutions. As the papers in Schwartz and Nichols (2006) describe, it is the nature of the society and the integration of its institutions and the exploitation of the environment that determine aspects of recovery from collapse. In this perspective, the ability or failure of institutions to provide for adaptations to stress that is an essential quality. My first book (Caldararo, 2004) was on the topic of sustainability as I was disturbed by the constant reference to "sustainability" without any practical definition, so I developed one based on a survey of human societies where ethics, guiding practice (whether religious or imposed by conditions), created limits to growth. Han Carl von Carlowitz, a Saxon tax accountant and mining administrator (24 December 1645-3 March 1714) is often referred to as the inventor of “sustainable yield forestry” (Schmithusen, 2013). As most indigenous people do practice it as Betty Meggers (1970) has so expertly detailed, von Carlowitz recognized its value before deforestation by industrial use became a crisis.

Environmental determinism has had a long history even if one only begins with the comprehensive treatments by Ratzel (1896-98). It was offered as a central agency in anthropology in America by Clark Wissler (1926) and developed by some of his students and associates (Dixon, 1928). Other theoretical questions concerning specific regions were addressed by researchers in the post World War II period as in that by Schwartz (1957). More recent efforts have attempted to utilize modern techniques to assess social responses in a variety of regions and cultures (Hastrup, et al. 2009). Speculation from the archaeological record has been made in the recent past, the first thorough analysis by Bell (1971). At about the same time Frank (1978) began a long series of works attempting to describe cycles of accumulation and dispersal of wealth driven by a number of factors including climate that still requires precise characterization. Both were preceded by Carpenter’s 1966 study of European civilizations and their responses to what he considered to be significant climate change in the past. Morato, et al. (1978) applied archaeological techniques to interpret patterns in California related human ecology, investigating cycles of weather still of use in understanding contemporary drought conditions.

Information in archaeological reports has often been collected and interpreted as products of general theoretical propositions (Caldararo, 1984). Nevertheless, archaeological investigation can provide sources of interpretation on how people lived and responded to climate change in the past. Megger’s 1971 study is a classic example. Scientific interest in the relationship between human ecology and changing patterns of the distribution of plants and animals has a long history (Thomas, Sauer, Bates & Mumford, 1956). Determining specific patterns of exploitation and human agency has depended on the development of new techniques (Caldararo, 2002; 2004), yet often interpretation is problematic as in the identification of carbon in strata and anthropogenic fire (Goldberg, 1985).

Reviews (e.g., Nature, 8 February 2007) of the Intergovernmental Panel on Climate Change (IPCC) first clearly reported that climate change is "Very likely" due to human agency and Crowley (2000) has shown that climate warming factors of pre-anthropogenic decadal-scale are unlikely to explain present warming. The 5th Assessment Report (AR5) strongly reinforces this view. In March of 2013 two articles appeared demonstrating considerable support for this conclusion. One study tracked the spread of plant life into colder zones (VanDerWal, Murphy Kutt, et al. 2013) and another reported on studies of ice and sediment cores to calculate temperatures in an unbroken record over the past 11,300 years (Marcott, Shakun, Clark & Mix, 2013).

## Ethnohistory and Climate Science

The role of societal collapse based on environmental degradation and anthropogenic driven climate change in the past has been proposed by a number of books in the last four decades and a recent review (Tainter, 2008) questioned the robust nature of these theories. Keeping in mind the paleoclimatological background (Hays, Imbrie & Shackleton, 1976) it seems obvious that human input to climate change has been considerable as population has increased and technological demands on carbon-based fuels and especially non-renewable fossil fuels has dramatically accelerated. Yet an editorial in Nature (2007) notes the field of research into climate change demonstrates some degree of uncertainty. Some uncertainty can be identified in the process of manufacturing models (Roscoe, 2014), while some results from how past climate in human ecological settings is interpreted. For example, William F. Ruddiman (2004; 2005) has argued convincingly that human farming in the Neolithic produced significant greenhouse gases to affect the planet's climate. Not everyone agrees with this view, and Schmidt (2005) has criticized some of Ruddiman's interpretations of the data. Williams (2006) pictures a slightly different scenario, but consonant with Ruddiman's general scenario. Taylor, et al., (1997) have shown how a significant change in climate can take place in a relatively short period of time, Holocene – Yonger Dryas Transition – over a 40 year period in 5 year steps driven by not a single factor, e.g., volcanism, but rather on a number of coincident changes leading to atmospheric modifications in water vapor. Chandler (1996) commented on how the reduction in solar radiation 115,000 years ago could not in it self be responsible for maintenance of year-round snow cover of the glacial period that followed, and suggested several contributing processes including tundra expansion.

Combination of environmental factors and climate change have likewise contributed to the collapse of human civilizations, especially where human activities have made societies, as Caldararo notes (2004) less resilient and degraded the ecological resources, consistent with cycles of expansion and reduction of extent and complexity (Braudel, 2001; Algaze, 1989). Tainter (1988) argues a more complex combination of factors, but ultimately it is a systems theory where he sees complexity as expensive and inputs ("investments") as having diminishing returns. All complex societies, therefore, in Tainter's view are limited systems fragile in that the diminishing returns are often offset by outside challenges that overwhelm the ability of the system to respond. In many ways this approach is reminiscent of Bell's (1971). However, as Sanders (1974) has shown for the evolution of states in ancient central Mexico, many civilizations develop within ecological settings that provide substantial challenges for sustainable food production and yet, in a pattern of Toynbee's challenge and response hypothesis (1934-1961), the cultural foundations of the social structures that evolve are adapted to severe variations and can respond displaying substantial plasticity. We can see this in the ethnological literature as in Barth's (1961) analysis of Iranian nomads where the Basseri are less able to adapt to pressures from the government than the Yomut and Komachi who were able to maintain social cohesion longer due to redistributive traditions the Basseri lacked.

With this scenario in mind we might consider the fact that studies of some ant societies have found they are able to ignore the environmental changes in their ecological systems due to their society complexity. Parr and Bishop (2022) have reported for one group, Hymenoptera formicidae, that it appears their eusocial adaptation and its attendant institutions, allows it to avoid the stresses of climate change. This fact may have important implications for human society, as we are highly structured, eusocial animals at this time (Caldararo, 2017). While Parr and Bishop (2022) consider a number of factors impinging on ant physiology and social life that can be interpreted as allowing them to adapt to environmental threats due to climate change, one can also imagine that this shielding of the species by its social life can be both a benefit and a danger as it may prevent needed responses to develop adequate modifications for survival. The same could be true for humans, we become so enclosed by the "hum of the hive" of our superorganism (Caldararo, 2020) of social life, that our trust in our technological abilities and history of dominance of the Earth blinds us to danger and prevents timely change in our behavior.

## Results: The Khmer Context

Some analysis of climate data, for example, Yancheva, et al., (2006) have produced a useful record of climatic change in East Asia but their claim that these changes were sufficient to cause civilizations to collapse is premature. What is contradictory to their argument is the history of the Khmer whose architecture, and empire rises before the onset of their identified dry periods and ends long after the last dry period terminates (Higham, 2001; Mabbett & Chandler, 1995; Pym, 1960), from before 400 A.D with the appearance of the Funan kingdom. to the mid 15th century. In fact, Briggs (1951) argues a different scenario where the Kambuja or Angkor Period extended from A.D. 802 to the capture of Angkor by the Siamese in 1431 and the definitive removal of the capital to the southeastern part of the kingdom in 1432. During all this period, the capital was at, or near, Angkor. This was the Classical Period of Cambodian history-the period of the far-flung empire, of the marvelous works of architecture and art, of the wonderful Sanskrit inscriptions.

But recent archaeological work has placed this view in question, discounting the ethnohistorical record as associated with specific special interests of the states the writers were reporting for and privileged in their interpretations. A three part relationship is constructed from this new evidence creating an original state, Funan, a successor, Angkor and with its failure, a Khmer Empire (Stark, 2006).

This later Khmer empire's collapse is often attributed to disease (e.g., malaria, see Welch, 1998) but a study by a Chinese diplomat/social scientist, Chou Ta-Kuan in the 13th century reports on a society in orderly condition. Agriculture is observed to be continuing without special stress, social order is apparent and food stuffs are plenty. Other researchers than Yancheva, et al. 2006, cite the possibility of a failure of the irrigation system (Buckley, et al., 2010), but there is no evidence of this in the eye-witness reports of Chou Ta-kuan, Ma Tuan-lin or of Arab and Indian merchants and visitors. This may give us an idea how people view significant changes and adapt to them as short term problems. Complicating this is Stark's (2006) argument that serious decline began after 1250 A.D. long before the climate event.

Ghosh (1960) compared the early reports and his conclusions parallel those of Higham (2001) who decided that the decline may have been the result of a number of factors together weakening the state but that the final blow was the military threat from the Thai kingdoms that eventually sacked the city of Angkor in 1431. One could perhaps argue, as Bell (1971) did and Frank (1978) and Chew (2007) assert, that the climate change created regional instability and that while the Khmer were able to adjust their technology and behavior to the local changes affecting agriculture, the peoples around them were under increasing ecological stress. Others argue that as more archaeological work is done the decision to abandon the site will become clearer, but it seems to have been a combination of the result of human urbanism and agriculture as well as climate change in their estimation (Stone, 2006).

## Discussion: History of Collapse Studies

Some authors make reference to Jared Diamond's book *Collapse* (2005) which is a renewal of Ratzel's (1896-8) views, (derived from works like Vico's cycles (Vico, 1744) and those of Ibn Khaldun, Spengler, and Sima Qian in105 B.C.E. and others) along with a popularization of some more complex discussions. Joseph Tainter has taken these theories to task in his, *The Collapse of Complex Societies* (1988). He has also questions regarding the focus on human activities as having significant role in the collapse of civilizations, in a recent review he downplayed the interaction promoted by others (Tainter, 2008). Diamond in his *Guns Germs and Steel* (1997) has built his vision of civilization on the belief that development is the highest value of human achievement. He ignores the anthropological premise that there is no one path of human expression, that industry, food production and masses of people are not necessarily positive values or ends.

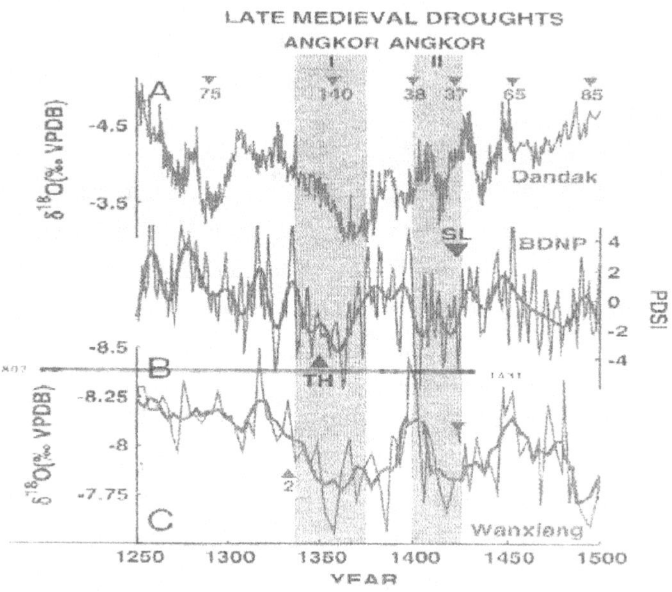


**Chart 2 Adapted from Buckley, et al., 2010. The green line shows the period of greatest cultural development of Khmer culture according to Briggs (1951).**

That the Australian Aborigines in 50,000 years do not produce an Einstein and a Theory of Relativity or build a machine because this might not be a necessary outcome of human life seems lost to him. To make up for what seems a deficiency compared to a progressive ideal of civilization, he returns to environmental determinism. That civilization is not necessarily a special achievement but rather a form of social complexity reached also by some social insects some hundred million years ago (Danforth, 2002). This knowledge should allow us to assess complexity and progress as any behavioral response of animals.

What Tainter (1988; 2008) shows is that societies have numerous components that allow them to sustain substantial losses in various spheres, economic, political and climatic, and yet survive and even expand. On the other hand, if these components are lacking or degraded due to internal conflict, mismanagement of resources, etc. they will be less resilient and will collapse or contract. Tainter (1988) does not place as positive an emphasis on the allocation of energy in societies in the production of prestige goods and their consequences as Shepard B.Clough (1961) does. Brian Fagan's (2000) position is more in support of Clough. However, Willey & Shimkini's work (1974) supports Tainter's argument, they note that, like the Khmers, Mayan society experienced a substantial population growth before collapse; this put pressure on the food production capacity of the land, rising class distinctions and socio-political conflict and competition set the stage for collapse. Scarborough and Burnside (2010) extend this parallel from Mayan history to Bali. Webster (2002) argues that this population expansion left the Maya particularly vulnerable to droughts that occurred later. This could be the case with the Khmer, but mediated through outside pressures; still growth of population does not necessarily indicate social stress (Hutchinson, 1967), but the opposite, positive conditions for reproduction and especially food supplies.

Marcus (1989, 1992, 1993, 1998) has described a cyclical pattern of collapse and renewal for the Mesoamerican pre-Columbian states, termed the "dynamic model." Environmental degradation by overpopulation during a growth cycle which coincided with a climate challenge would accelerated collapse. This parallels other deterministic models.

Debate on a correlation of food supplies and population has varied as Hutchinson (1967) has shown. A key element is social discipline and the symbols that maintain social solidarity, a factor examined in detail by Blackman (1961). Nevertheless, the Khmer flourished when Buckley, et al. (2010) argue they should have been in decline. One has to note, however, that the period between the low points in 50 year intervals shows a considerable rebound in rainfall and though the authors interpret additions to the Khmer water system during this time and after, even this indicates organization and response. Nevertheless, Khmer collapse takes place after a period of monumental building as in the Mayan case and as is seen at the end of the Sixth Dynasty in Egypt. Reorganization and renewal take place later among the Khmer but further downstream (Coe, 2003).

Moundbuilder societies present a unique problem as Abrams (2009) notes:

> "The archaeological expectations of a collapse do not appear for Hopewellian societies. Neither evidence of widespread diseases nor heightened death rates have been recovered (Dunnell and Greenlee, 1999). Similarly, the data for environmental decline are limited although the warming episode c. A.D. 400 of (Baerreis et al., 1976; Loehle, 2007) encourages a greater investigation of dynamic environmental ends."

In their paper Buckley, et al. (2010) argue that substantial droughts contributed to the Khmer collapse as they describe two periods, one in the early 13$^{th}$ century and another in the early 14$^{th}$. However, Khmer methods of agriculture seem to have been less a part the failure to adapt in the region rather it may have been the depletion of resources to building prestige architecture and regional conflict. This may be an area of consonance between several of the theories, of Tainter, Chew, Marcus and Diamond and Caldararo. Here the inputs are too expensive to maintain stability, rather they contribute to increasing inequality and degradation of social capital and internal solidarity that weaken a society and make it victim to external challenge. Therefore, the very symbols of civilization can be seen as costs and elements of social stress that explains why such investments might be timed with exterior stresses as an attempt at compensation that only makes the situation worse by expending scarce resources, as in Diamond's focus (1997) on the Bird Man cult in Easter Island, though its collapse has been recently association with European contact.. I discuss the economic implications of these factors in more detail in a book (Caldararo, 2014).

## Conclusion

While climate must be seen to play a role in the development and demise of civilizations, no single factor usually stands out although single cause theories abound in the history of social science and history, the story is often more complex and frustrating. What is clear is that many societies note environmental stress when it takes place in unusual intensity or patterns, but are often unable to respond sufficiently to compensate within the context of their available resources or by modifying their values and cosmology that are often the driving forces behind resource depletion. In his time the Roman historian Polybius, noted that recognition of degradation and having the power to change are quite different aspects of society. This seems especially true today of the global economy and the attitudes denying climate change as caused by human activity. Many societies deny the link; few escape the results, and most suffer the consequences when prudent action might have averted disaster.

What is undeniable, however, as Professor Sing C. Chew has shown (2007) is that "... human induced changes to the ecology and the climate in turn form barriers to the reproduction" of the natural capital of the world system. One might go further in geological history and note that animal societies are adaptations to environmental changes not under the control of the species, whether cyclical like seasonality or periodic like changes in solar energy due to planetary tilt, reduced radiation or other factors. Insect societies are complex (like bees, wasps and ants) and have evolved to sustain their members over weather conditions where food resources are low and to produce cooperative patterns to exploit food resources when conditions allow. They also have behavior patterns that provide structures to protect them from unusual environmental conditions to a certain extent. Human society has been much like these earlier examples, especially ants and bees. Human society is at the point of reflexivity where it has become aware of itself and has the potential to not only moderate the effects of its own physical presence, but to interfere with the global environment to produce a more comprehensive sustainable adaptation.

Teilhard de Chardin (1959) believed that there was an inherent trend in the evolution of life of intelligence and consciousness, and one might think so given the fossil record, especially since the Pre-Cambrian development of complex organisms. Yet it depends on how one defines the terms, fungi in the tens of thousands and millions seem to produce together kinds of intelligence and communication (Hathaway & Arevalo, 2023). Some species cover acres functioning as one individual, perhaps like a human brain (Wu, 2022), and they seem to be benefiting from our destruction of the planet. Whether what results can be considered consciousness is perhaps still a debate for philosophers. But our species may not be the apex of this evolution of consciousness and intelligence.

Chew (2007) argues that collapse and "Dark Ages" have a renewal function given the degradation of the environment resulting from human accumulation of wealth in all its forms, both physical and symbolic. Caldararo has shown elsewhere (2004; 2009), however, sustainability and stability are possible without collapse dependent on cultural proscriptions against certain levels of accumulation that would lead to degradation and collapse. The Khmer appear to be an example of this variation. The missing element to the other theories of collapse is their failure to account for this variation, especially the role of ideologies. Rostovtzeff (1926) suggests that Neo-Pythagoreanism was at the heart of the Roman economic expansion after the Civil Wars and was a central element in rebuilding the stability of the Empire. Reischauer & Fairbank (1960) pursue a similar role for Neo-Confucianism in the dynastic form in China and I have argued that a culture that has integrated Shintu, Neo-Confucianism and Buddhism has allowed Japan to produce a resilient response to challenge as well (Caldararo, 2003). The Japanese have responded to environmental stress that reduced resource availability (e.g., food) several times by the reduction of population (Fisher and Sargent, 1977). The country is undergoing a current period of population decline which could reduce its resource extraction to a closer frame of carrying capacity as in the past only with now industrial technology. The rest of the world's current economic system is based on an ideology of capitalism requiring constant growth and consumption and its promise of riches, but it functions on magical thinking and irrationality that also promotes inequality and environmental degradation as Caldararo has shown in a number of publications recently (Caldararo, 2004; 2009; 2012).

Regeneration after collapse is often affected by climate, especially in some cases as in Mayan patterns (Marcus, 1998). Regeneration takes a variety of forms dependent on local environment, extent of collapse, intrusive peoples and type of complexity of the original society and its integration (Blanton, et al. 1996). Stark (2006) has emphasized the idea of regeneration and movement in the Khmer example with the idea the continuity of symbols and institutions and an Indic ideology allowed for success. The failure of regeneration in the case of the influx of a powerful new culture and society coupled with efforts to prevent continuity of native symbols and institutions can be seen in the case of Native American complex societies where destruction of Native architecture, writing, learning and institutions of nearly every type created a permanent hiatus concerning Native complex culture (Caldararo, 1994).

While it is obvious from the fossil record that climate change has occurred in the past, the factors involved, reduced solar radiation, Milankovitch cycles of axis variation in tilt, albedo effect and the influence of life processes, are many. It is clear, nevertheless, that life processes have an effect. The Great Oxidation Event that occurred about 2.3 billion years ago was the result of life processes. Cyanobacteria had been pumping out oxygen for more than a billion years and apparently this accelerated between 2.4 and 2.3 billion years ago dramatically changing the atmosphere of the earth (Margulis, 1981; see Tomkins, et al., 2016 for an alternative view) and clearly toxic conditions to anaerobic forms.

Loss of diversity at the end of the Devonian has been explained by a reduction in speciation and invasions of cosmopolitan species, rather than any single event, but each extinction period is associated with anoxic sea conditions that may be the result of volcanic action or biological activity by certain species and their metabolites (Lyons, et al., 2009). While other factors may play more important roles in certain extinctions and changes in climate, for example possible asteroids in the Permian extinction or at the end of the Cretaceous, it is clear that life processes do have a role in climate change (Wignall and Twitchett, 1996).

The question for people today is one of consciousness and what it means to be human. We often pride ourselves by believing we are thinking animals, so one might expect that given past roles "dumb" animals and plants have played in climate change, should not we recognize our own role and behave accordingly? Will we create conditions that are toxic to our own survival as other life forms have done in the past? Sagan and Shklovskii (1966) thought that this was a likely outcome, but mainly due to the short duration associated with intelligent life in their theoretical framework. However, Cai, et al., (2020) come to a similar conclusion and the evidence presented in their paper is as convincing.

## References

Abadi, Mehrdad Sardar, Jeremy D. Owens, Xiaolei Liu, Theodore R. Them, II, Xingqian Cui, et al., (2019) "Atmospheric dust stimulated marine primary productivity during Earth's penultimate icehouse," *Geology*, 17 Dec. doi: 10.1130/G46977.1

Abrams, Elliot M. "Hopewell Archaeology: A View from the Northern Woodlands," *Journal of Archaeological Research*, Vol. 17, No. 2, 2009, pp. 169-204.

Algaze, Guillermo "The Uruk expansion: cross-cultural exchange in Early Mesopotamian Civilization," *Current Anthropology*, v. 30, n. 5, Dec.,1989:571-608.

Barth, Fredrik (1961) *Nomads of South Persia*, The Little, Brown Series in Anthropology, Boston: Little, Brown & Co.

Bell, Barbara "The Dark Ages in ancient History," *American Journal of Archaeology*, v. 75, n. 1, 1971: 1-26.

Blackman, H.J. *Political Discipline in A Free Society*, London: G. Allen &Unwin,1961.

Blanton, R.E., G. M. Feinman, S.A. Kowalewski, and P.N. Peregrine, "A dual-processual theory for the evolution of Mesoamerican civilization," *Current Anthropology*, v. 37,1996:1-14.

Braudel, Fernand (2001) *Memory and the Mediterranean*, ed., Rosclyne De Ayala and Paule Braudel, trans. Sian Reynolds, New York: Vintage Books.

Briggs, Lawrence Palmer "The Ancient Khmer Empire," *Transactions of the American Philosophical Society*, New Series, V. 41, 1, 1951: 1-295.

Buckley, Brendan M., Anchukaitis, Kevin, J., et al., "Climate as a contributing factor in the demise of Angkor, Cambodia," *PNAS*, v. 107, v. (15), March 27, 2010: 6748-6752.

Cai, Xiang, Jonathan H. Jiang, Kristen A. Fahy and Yuk L. Yung, (2021) "Statistical estimation of the occurrence of extraterrestrial intelligence in the Milky Way Galaxy," *Galaxies*, v. 9, n. 1, https://doi.org/10.48550/arXiv.2012.07902.

Caldararo, Niccolo (1984) "A note on the archaeological report," North American Archaeologist, v. 6, n. 1: 63-80.

Caldararo, Niccolo (1994) "The lost libraries of the Americas," Library History, v. 10,: 9-18.

Caldararo, Niccolo (2002) "Human ecological intervention and the role of forest fires in human ecology," Science of the Total Environment, v. 292: 141-165.

Caldararo, Niccolo (2003) "Japan and the concept of Sustainability," International Journal of Anthropology, v. 19, n. 3: 213-230.

Caldararo, Niccolo (2004) *Sustainability, Human Ecology and the Collapse of Complex Societies*, Lewiston: Edwin Mellen Press.

Caldararo, Niccolo (2009) "Derivatives and debt: The ␣ market as god and marketing as proselytizing," chapter 6 in Charles V. Karsone editor, Finance and Banking Developments, New York: Nova Science Publishing.

Caldararo, Niccolo (2012) *The Anthropology of the Credit Crisis, Irrationality, Magical Thinking and the Role of Inequality*, San Francisco: MTC Press.

Caldararo, Niccolo (2014) *The Anthropology of Complex Economic Systems*, Lanham: Lexington Books.

Caldararo, Niccolo (2020). "The Hum of the Hive." *Anthropology News* website, September 10, 2020. DOI: 10.14506/AN.1482.

Carpenter, Rhys (1966) *Discontinuity in Greek Civilization*, Cambridge: Cambridge University Press.

Chandler, Mark "Trees retreat and ice advances," Science, v. 381, 6 June, 1996:477-8.

Clough, Shepard B. *The Rise and Fall of Civilization,* New York: Columbia University Press, 1961.

Chew, Sing C. (2007) The Recurring Dark Ages, Lanham: Rowman and Littlefield.

Chou, Ta-kuan (1951) *Memoires sur les coutumes du Cambodge de Tcheou Takouan*, Trans. P. Pelliot, Paris: Ecole francaise d'Extreme-Orient.

Coe, Michael D. (2003*) Angkor and the Khmer Civilization*, London: Thames and Hudson.

Crowley, Thomas J. "Causes of climate change over the past 1000 years," *Science*, v. 289, 14 July, 2000:270-5.

Danforth, B.N. "Evolution of sociality in a primitively eusocial lineage of bees," *PNAS*, Jan 8, (2002): 286-290.

Diamond, Jared (1997) *Guns, Germs and Steel*, New York: W.W. Norton.

Diamond, Jared (2005) *Collapse,* London: Penguin.

Dixon, Roland B. (1928) *The Building of Cultures,* New York: Charles Scribner's Sons.

Dubos, Rene, (1973) "Humanizing the Earth," *Science* 179, Feb 23, 4075. pp. 769-772.

Dubos, Rene (1974) "The despairing optimist," *The American Scholar*, Spring, pp. 192-3.

Dubos, Rene (1970) *Reason Awake*, New York, Columbia University Press.

Editorial, 2007, "What we don't know about climate change," *Nature,* v. 445, 8 Feb. 580-1.

Fauci, Anthony (2024) *On Call: A Doctor's Journey in Public Service,* New York, Penguin.

Fisher, Charles A. and John Sargent, (1977) "Japan's ecological crisis," in *Global Perspectives on Ecology,* Thomas C. Emmel with Lieselotte Hofmann (eds) Palo Alto, Mayfield, 295-309.

Fagan, Brian (2000) *Floods, Famines and Emperors: El Nino and the Fate of Civilizations*, London: Pimlico.

Frank, A.G. (1978) *World Accumulation 1492-1789*, New York, Monthly /Review Press, London: Macmillan, 1978.

Ghosh, Manomohan (1960) *A History of Cambodia*, Saigon: J.K. Gupta.

Goldberg, F.D., (1985) *Black Carbon in the Environment*, New York: J. Wiley.

Grimes, David Robert (2020) "Health disinformation and social media," *EMBO reports,* Nov. 5, v. 21, n. 11, e51819.

Hastrup, Kirsten ed., (2009) *The Question of Resilience: Social Responses to Climate,* Copenhagen: Det Kongelige Danske Videnskalbernes Selskab.

Hathaway, Michael & Willoughby Arevalo (2023) "How do fungi communicate," MIT Technology Review, April 24.

Hathaway, Michael (2022) *What a Mushroom Lives For,* Princeton, Princetown University Press.

Hays, J.D. John Imbrie, N.J. Shackleton, "Variations in the earth's orbit: pacemaker of the Ice Ages," *Science*, v. 194, n. 4270, 10 December, 1976:1121-1132.

Higham, Charles (2001) *The Civilization of Angkor*, Berkeley: University of California Press.

Hutchinson, E,P. (1967) *The Population Debate*, NewYork: Houghton Mifflin.

Katella, Kathy (2024) "3 things to know about XEC, the latest Covid strain," Yale Medicine, Oct. 18.

Lofgren, Mike (2022) "The rise of the right-wing university," *Common Dreams,* Oct. 3rd.

Lyons, Timothy W., Ariel D. Anbar, Silke Severmann, Clint Scott and Benjamin C. Gill, (2009) "Traking euxinia in the ancient ocean: a emultiproxy perspective and Proterozoic case study," *Annual Review of Earth and Planetary Sciences,* v. 37, n. 1, 507-553.

Ma, Tuan-lin (1976*) Ethnographie des peoples etrangers a la Chine, ouvrage compose ...par Ma Touan-lin,* Trans. Le Marquis d'Hervey Saint-Denys, Paris: Georg.

Marcott, S.A., J.D. Shakun, P.U. Clark, A.C. Mix, "Reconstruction of regional and global temperatures for the past 11,300 years," *Science*, 339, 6124, 2013:1198.

Marcus, J. "From centralized systems to city-ststes: possible models for the Epiclassic," in R. Diehl & J.C. Berlo, eds. Mesoamerica after the Decline of Teotihuacan, AD. 700 – 900, Washington, D.C.: Dumbarton Oaks, 1989:201-8.

Marcus, J., "Dynamic cycles of Mesoamerican states," National Geographic Research and Exploration, v. 8, 1992: 392-411.

Marcus, J. "Ancient Maya political organization," in J.A. Sabloff & H. Henderson, eds., Lowland Maya Civilization in the Eight Century, Washington, D.C.: Dumbarton Oaks, 1993:111-184.

Marcus, J. "The peaks and valleys of ancient states, an extension of the Dynamic Model," in G. Feinman & J. Marcus, eds., Archaic States, Santa Fe: School of American Research, 1998:59-94.

Mabbett, Ian & David Chandler, The Khmers, London: Blackwell, 1995.

Margulis, Lynn (1981) *Symbiosis in Cell Evolution: Life and its Environment on the Early Earth*, San Francisco: W.H. Freeman & Company (originally published in 1971).

Meggers, Betty (1996) *Amazonia: Man and Culture in a Counterfeit Paradise* (revised Edition), Washington, D.C., Smithsonian Institution.

Moratto, Michael J., Thomas F. King, Wallace B.Woolfenden, "Archaeology and California's climate," The Journal of California Anthropology, Vol. 5, No. 2 WINTER 1978: 147-161.

Parr, Catherine, L. and Tom R. Bishop, (2022) "The response of ants to climate change," *Global Change Biology*, v. 28, n 10: 3188-3205, 11 March, doi.org/10.1111/gcb.16140.

Pierre, J.M. (2023) "Conspiracy theory belief: a sane response to an insane world?" *Review of Philosophy and Psychology*, doi.org/10.1007/s13164-023-00716-7.

Pym, C. (1960) *The Ancient Civilization of Angkor*, London: Hodder & Stoughton.

Ratzel, Frederick (1896-8) *The History of Mankind*, Translated from the second German edition, London: Macmillan.

Reischauer, E.O. &, J.K. Fairbank, East Asia: The Great Tradition, Boston: Houghton Mifflin, 2010.

Roscoe, Paul "A changing climate for anthropological and archaeological research? Improving the climate-change model," American Anthropologist, v. 116, n 3, 2014: 535-548.

Rostovtzeff, M.I., (1926) *Social and Economic History of the Roman Empire*, Oxford: Oxford University Press.

Ruddiman, W.F. "How did humans first alter global climate?," *Sci. Amer.,* March, 2005:46-53.

Ruddiman, W.F. (2005) *Plows Plagues and Petroleum: How Humans Took Control of Climate*, Princeton: Princeton U. Press.

Sagan, Carl and I. Shklovskii (1966) *Intelligent Life in the Universe,* San Francisco, Holden-Day.

Sanders, W.T., "Hydraulic agriculture, economic symbiosis, and the evolution of states in Central Mexico," in Lamberg-Karlosky, C.C. & Saloff, J.A., eds., The Rise and Fall of Civilizations: Modern Archaeological Approaches to Ancient Cultures, Cummings Menlo Park, 1974:119-134.

Scarborough, Vernon L. and William R. Burnside, "Complexity and sustainability: perspectives from the ancient Maya and the modern Balinese," American Antiquity, Vol. 75, No. 2 April, 2010: 327-363.

Schmidt, Gavin I., "Methane: A scientific journey from obscurity to climate super-stardom," Goddard Inst. For Space Studies, New York, N.Y, 2005.

Schmithusen, Franz Josef (2013) *Three Hundred Years of Applied Sustainable Forestry*, Working Papers, Forest Policy and Forest Economics Department of Forest Sciences International Series, EWTH, Swiss Institute of Technology, Zurich.

Schwartz, Douglas W. "Climate change and culture history in the Grand Canyon Region, American Anthropologist, v. 22, n. 4, 1957: 372-377.

Schwartz, Glenn M. and John J. Nichols, (eds.) 2006 After Collapse: The Regeneration of Complex Societies, Tucson, University of Arizona Press.

Shephard, Lauren Lassabe (2023) "The history behind the Right's effort to take over American universities," *Time Magazine,* Oct 23.

Stark, Miriam T. "From Funan to Angkor: Collapse and Regeneration in Ancient Cambodia," in Glenn M. Schwartz and John J. Nichols, eds. After Collapse: The Regeneration of Complex Societies, Tucson, University of Arizona Press, 2006:144-167.

Stone, Richard "End of Angkor," Science, New Series, Vol. 311, No. 5766, Mar. 10, 2006: 1364-1368.

Tainter, Joseph, The Collapse of Complex Societies, Cambridge: Cambridge University Press, 1988.

Tainter, Joseph, "Collapse, sustainability and the environment: how authors choose to fail or succeed," Reviews in Anthropology, v. 37, 2008:342-371.

Taylor, K.C., P.A. Mayewski, R.B Alley, et al., "The Holocene – Younger Dryas Transition recorded at Summit, Greenland," Science, v. 278, 31 October, 1997: 825-7.

Teilhard de Chardin, Pierre (1959) The Phenomenon of Man, London, Collins & Sons.

Thomas, Jr., William, L., Carl O. Sauer, Marson Bates, and Lewis Mumford, Man's Role in *Changing the Face of the Earth,* 2 vol.s, Chicago: University of Chicago Press, 1956.

Tomkins, Andrew G., Lara Bowit, Matthew Genge, Siobahan A. Wilson, Helen E.A. Brand et al., (2016) "Ancient micrometerorites suggestive of an oxygen-rich Archaean upper atmosphere," *Nature*, v. 533, n. 7602, doi:10.1038/nature17678.

Toynbee, Arnold 1934-1961, *A Study of History*, Oxford: Oxford University Press.

VanDerWal, Jeremy, Helen T. Murphy, Alex S. Kutt, et al., "Focus on poleward shifts in species' distribution underestimates the fingerprint of climate change," *Nature Climate Change*, March v. 3, n5, 2013. 239-243.

Vico, Giambatista, (1744) *The New Science*, trans. By Thomas Goddard Bergin and Max Harold Fisch, 3rd edition, 1968, Ithaca, Cornell University Press.

Webster, D. (2022) *The Fall of the Ancient Maya*, London: Thames & Hudson.

Welch, David J. "Archaeology of Northeast Thailand in Relation to the Pre-Khmer and Khmer Historical Records," International Journal of Historical Archaeology, Vol. 2, No. 3, *The Transition to History in Southeast Asia Part I: Cambodia and Thailand*, 1998: 205-233.

Wignall, Paul B. Richard J. Twichell (1996) "Ocean anoxia and the end Permian mass extinction," *Science*, v. 272, n. 5265, 1155-1158.

Willey, Gordon R. & Demitri B. Shimkin, "The collapse of Classic Maya civilization in the Southern Lowlands: A Symposium Summary Statement," in Lamberg-Karlosky, C.C. & Saloff, J.A., eds., *The Rise and Fall of Civilizations: Modern Archaeological Approaches to Ancient Cultures*, Cummings Menlo Park, 1974:96-103.

Williams, Michael, *Deforesting the Earth, from Prehistory to Global Crisis,* Chicago: University of Chicago Press (2006).

Wissler, Clark (1926) *The Relation of Nature to Man in Aboriginal America,* New York: Oxford University Press.

Wu, Katherine (2022) "The bigger this fungus gets, the worse we're doing," *The Atlantic,* August, 15.

Yancheva, Gergana et al., "Influence of the intertropical convergence zone on the East Asian monsoon," *Nature*, v. 445, 2006:74-77.

# Chapter Seven

# Destroying Institutions: Attack on the EPA, Abandoning the Medical Profession During a Pandemic

## Abstract

Institutions have been described as essential patterns of behavior that come to support a society and be generally understood and agreed upon in their roles. They can include behaviors that individuals must participate in as in rites of passage in religion, or in the use of money or joining a military or providing labor for communal structures or means as in irrigation. When one segment of a society comes to disagree with the goals of institutions and withdraws support or actively attacks one or more, it threatens not only the existence of the institution but the society as an organization. Often in history revolutions are built on such attacks as in those of the Optimates in ancient Rome or the Bolsheviks in Russia. Today a political movement supporting former President Trump has attacked institutions of justice (courts, Justice Department), security (FBI, CIA), military, education, news, elections and the medical profession. The attack on the medical profession has been one that has a longer history, one bracketed in disbelief in science in general and has set the two institutions of religion and medicine at odds. An underfunded medical profession has faced the Covid-19 pandemic in a cloud of conspiracy theories that question even the purpose or need for medicine. The consequences during Covid-19 have been catastrophic by many assessments. One can see from the history of the Trump administration's attack on the Environmental Protection Agency and other government programs that active attacks on institutions that could reduce the effects of climate change are an essential problem of human adaptation.

Key Words: Institutions, Covid-19, revolution, medicine, science, renewal, anti-science attitudes, news, elections
JEL Classification Codes: I0; I1; I2; I3; Z1

## Introduction

History has shown that societies often reach points of disorganization where they are unable to maintain order and function (Tainter, 1988). Usually some failure in core institutions on which a people depend for essential services or reinforcement of basic trust in reciprocation of wealth and labor, leads to degradation of social credit and agreement in the social contract of daily life. Often degradation can continue over a long period of time depending on internal traditions and custom and if external forces perpetuate the appearance and hope of stability, as in Lebanon over the past 100 years since the fall of the Ottoman Empire. American institutions are under assault today and these, including law, journalism and the news, political order, medicine and elections, have become the target of elements in society determined to undermine the current American social contract.

Looking at social groups under stress and analyzing the function of institutions and changing attitudes of a population or segments of the society to them, is one way of understanding functional questions of integration. Geertz (1973) criticized the failure of social science to produce new, more useful theories of social action than those of the early 20th century. He argued that while Parsons (Parsons & Shils, 1951) had reinvigorated the useful ideas of Weber, he notes that Malinowski's (1948) work was still of value. Yet in the post-World War II crisis of colonial rule and economic disruption it was, in my opinion, Malinowski's later work that provides the most workable view of the challenges (1945). This is a functionalist approach that Malinowski's students applied in the pre-World War II period and has been revived since Geertz's comments.

Geertz (1973) reviews many of the problems facing modernization and post-colonial economics, and he realizes that adaptations to Western economic and political systems will have consequences that are fundamental to new secure and stable nation systems. His review of political institutions from Bali to Morocco, provided a firm perspective of the crises to come. In this chapter I look at how the effects of globalization have affected not only economic and political systems, but medical response as the medical institutions of the world have come under increased attack during the Covid-10 pandemic. This was magnified by the forces unleashed by earlier revitalization movements both in the USA (Reagan/Gingrich "Contract with America") and more recent European populist/nativist movements like Brexit (see Caldararo, 2022).

## Examples of Stress

In 2007 the financial industry caused the world economy to implode. Governments around the world came to the rescue of banks and investors, hedge funds and insurance companies. All these entities had had some responsibility for the crisis. Over the past 60 years the legal profession has assaulted the medical industry with malpractice lawsuits that are claimed to improve the responsibility of medical professionals, pharmaceutical firms and medical corporations and device makers to provide more effective service and products (Kessler, 2011). America's health care system is the most expensive in the world and seems to be less effective, and has been mired in inequalities (Starr, 1982).

While the financial crisis of 2007 did not place bankers in danger of losing their lives and a few did go to jail for fraud in the dotcom crash, as in the case of Enron's management, during the Covid-19 pandemic we are saw our medical professionals from doctors to orderlies risk their lives on a daily basis. People engage in jobs for a livelihood, often also for prestige, but unless they are in the military or police force they are seldom in harms way due to their work. It is strange in this regard that the most dangerous jobs in America are not police or military but include such work as forester and plumber, according to the U.S. Department of Labor Statistics.

It should be obvious by now that our entire economic system rests on a number of fallacies. John Maynard Keynes (1930) wrote at the beginning of the 20$^{th}$ century that by the end of the century most Americans would hardly have to work at all given the vast improvements in technology that were made. However, our society was based on an agricultural society concept of needs with little production of the tremendous kinds of consumer products we have today. Since the March 2020 lockdowns in

America and many other nations across the globe, millions of people have survived well without jobs and governments have produced cash for them to continue to purchase goods and services. Some people have faced economic constriction of their lifestyles, but in general we have been able to feed and clothe billions without work for hundreds of millions. What is lacking is a comprehensive relief act going beyond what governments have done so far as with deferring or modifying rents and mortgages, as well as extending unemployment insurance. Thus far most government support has not been made effective and loans to companies have mainly gone to large companies (Wiggleworth, et al., 2020). Concerns about debt are not supported by history, recovery is based more on liquidity as Jorda, et al., (2020) have shown. World debt grew over $1.3 trillion in 2024 to a record of $315 trillion 333% of GDP (Tiftik, et al,, 2024).

This situation brings up a fundamental issue of the goal of society and economies. If we can simply print money and still eat and be comfortable, it seems Keynes was right, but why then do we have to work so much? If the value of goods and services has not increased (inflation seemed conquered or at least tamed before the pandemic and continues to be so) then is there some basic contradiction that needs to be addressed? While debts are piling up in the formal sense of government expenditures over incomes, we are not experiencing stress in any of the financial markets. A small "correction" happened in March 2020 at the beginning of the pandemic, but was quickly eased by government actions. The nature of debt is largely theoretical and depends on how people view spending and future gains. American government credit for the Union during the Civil War fell dramatically and then rose as Union victories accumulated and tax revenues amounted to a fantastic level by the end of the war (Shultz and Caine, 1937). But the nature of credit and money is one that is irrational in a general sense as has described by Caldararo, (2014) and Grabner (2011) in a more concentrated work on the history of debt.

## Function of Institutions and Attacks

Societies create institutions to routinize and sanction provision for certain needs that are valued by the culture. Though there is some disagreement about what are specific institutional forms (Hodgen, 2006), there is a general understanding of their function and structure (Malinowski, 1945). When agreement in society is general and uniform expectations of the function of institutions and their cost to individuals is clear and accepted, they are successful, whether efficient or not by exogenous measure. Ideological explanations provide protection for institutions, and central figures with defined roles in societies often reinforce such explanations. There are examples of nations with pluralistic constitutions based on a variety of historic threats, as in Belgium or many of the former European colonies. Here institutions often have varied support or are divided so to perform for multiple cultural identities. Where these exist, integration of the variations and effectiveness of the institution is central (Hsu, 1969; Mair, 1969; Malinowski, 1945). Failure of agreement can result in change to institutions, both in the definition of their role and in modifications of explanations. Widespread dissatisfaction with how institutions work or provided benefits can result in their elimination or replacement. Forms of government, for example, constitutions are examples of this kind of modification or change. When there is widespread dissatisfaction with institutions in many areas of society or where groups within society make it their cause to reorganize society, then we can apply the word, "revolution" to such transformations if successful. Sometimes we have historical examples, as in ancient Athens and other Greek city-states and their attempts at government, of a back and forth adaptation of forms, one replacing the other in transitions that are peaceful or in violent overthrows (Larsen, 1955). Or the slow degradation of constitutional order devised and conceived of by a minority as a means to greater power and wealth as in Rome with the Optimates (Lintott, 1999; Morstein-Marx, 2004). This process of delegitimizing institutions is discussed in more detail in Pardo and Prato, (2019).

Since the rise of Republican Congressman Newt Gingrich, the political dialogue has been purposely degraded, becoming more bellicose and aggressive (Coppins, 2018). Much of this has been organized from the media, mainly talk radio. This flow of speech has been aimed at destroying traditional forms of agreement, as in what are trusted sources of information, what are means of verifying facts and how are biases in reports balanced or corrected. Following the 2020 election there began a series of attacks on election officials with the charge that the election was fraudulent, these have continued to this day (Halpern, 2021).

Attacks on institutions help divide the social narrative and create separate universes of knowledge about how events take place. One of the main vehicles of this process has been the production of conspiracy theories. As a genre of thinking and analysis, the idea of a conspiracy theory is not essentially different from any starting point in a theoretical engagement as in the scientific method. The essential difference is that the underlying motivation in the construction of a conspiracy theory departs fundamentally from the precepts of the Enlightenment in that instead of the search for truth or facts, the motive is to confuse, frustrate and cause people to respond to falsehoods as if they were facts. A means to this end has been to undermine education and create an atmosphere of bias in public education in general. Personal intent is the central characteristic, either to purposely distort reality to produce a desired effect in others, or as a defense mechanism for the disturbed individual and can be seen as a measure of dysfunction in a society (Pierre, 2023). As this has produced a "culture war" milieu where it is charged that colleges and universities are organs of liberal bias and indoctrination, areas of common ground have been disappearing in the national dialogue. These arguments ignore the fact that many Americans lost their jobs in the 20th century due to laws restricting speech, especially where teachers and other public employees were forced to sign loyalty oaths (Heale, 1998) or were fired due to their writing or speech. A fundamental focus has been since the 1970s on producing alternative education systems, in grades K to 12, colleges and especially subjects like law (Shephard, 2023; Lofgren, 2022). In the past year attacks on school boards and educators have increased (Thompson, 2021) creating a battlefield for the minds of children. The ultimate goal is to create a wed of distrust so complete that people who come to accept the basic ideas, then refuse to accept any contradictions to any part of the theory.

## The Attack on Medicine and Science

Today we are seeing a vast assault on medical science, undertaken in a comprehensive yet contradictory fashion where leaders deny disease and medical practice and advise their followers to also reject the idea of disease. Yet they take advantage of medical advances in curing Covid-19 as in the cases of President Trump and Rudy Giuliani. This attitude stands on a long history of anti-science that in America has created a movement against vaccination in general (Burki, 2020; Grimes, 2020).

While funding for medical research in the USA was minimal prior to WWII, funding ramped up substantially during the war and after. However, by 1990 a substantial portion of funding for medical search was from private sources, both non-profit foundations and corporations. Some concerns were raised about this trend regarding the influence of private sources on cost and availability of products as well as scientific independence of researchers (Bloom and Randolph, 1990). Medical education research, a different category where students learn how to do medical research has also been underfunded for years (Carline, 2004), producing a shortage of doctors and specialists some of which has been compensated for by importing foreign-trained MDs. USC's Keck School of Medicine reported before the impact of the Covid-19 pandemic struck in February of 2020, that there was a substantial shortage in trained medical staff with over 80,000 qualified applicants turned away from training programs in 2012 due to a lack of funding in training (Anon, 2020). The response of America's medical community in the face of the pandemic, given these deficiencies is remarkable. *Stat* reports that half of Amcrican medical facilities have a massive staffing shortage (Goldhill, 2020).

Fear of the disease has also caused many people to avoid hospitals and clinics, that has resulted in a 40% rise in excess deaths due to deferred treatment (Birkmeyer, et a., 2020; Valerio, 2020). But the main consequence of underfunding is eclipsed by the inability and wanton disregard of the Federal government and some state governments to address the pandemic as real and plan for the emergencies it has caused (Meyer and Madrigal, 2020). One cost to the USA and to other countries of the death rates is certainly more than just the lives lost, skills, impact on families, communities and loss of wealth in fighting the disease. But another hidden cost is the number of doctors, nurses and other staff that have fallen victim to the disease. More than 1,200 health care workers have been reported to have died of the disease (Rezba, 2020), yet more than 216,000 healthcare workers, mainly nurses have been found to be infected (Barrett, et al., 2020). The future effect of this on decisions of people to go into the field could be catastrophic.

The main cause of this situation is the anti-science attitude of the Republican Party and its president Donald Trump (Pazzanese, 2020; Tollefson, 2020). A result of this attitude and resulting political action is delays in resources to treat, study and develop new vaccines (especially for the growing number of variants) and treatments (Wood and Schulman, 2021). By mid 2020 several noted journals were calling for and end to political intervention in both the practice of medicine related to Covid-19 and to the development of vaccines, treatments and their application (Wakeman, 2020). The landscape for health care had never been so challenged (Fauci, 2024).

## Crime, Incomes, Mental Health and Education

It is an unexpected consequence of the pandemic that with the loss of jobs and income there has been so little unrest, violence or disruption. The PEW Research Center charted the loss of jobs among Americans (Parker, et al., 2020) with about half reporting no new jobs. The survival of families and individuals seems to have depended on the first Congressional allocations and unemployment insurance and well as efforts of charities and self-employment.

Yet it is curious how in general most crime statistics showed a decrease in crimes with the exception of shootings and killings tp 2020 (Corley, 2020). Yet when stay at home orders were lifted, a racial disparity appeared with a rise in crime in black majority neighborhoods (Harden and Jouvenal, 2020). It is also obvious that there has been an increase in domestic violence (Evans, 2020). Though from 2021 to mid 2022 crime rates jumped substantially by over 4% in come locations (Lopez, 2022).

The pandemic is causing a deficit in training and education with the greatest loss in those in the lower income scales and in people of color (Garcia and Weiss, 2020). It has also had a significant effect on the mental health of health care practitioners, hospital staff and clinics (Jacobs, 2021).

With the election of President Biden this situation has experienced some turn around that is having positive results, though a large portion of the country that supported Trump still avoids vaccination and is making up most of the new cases of Covid-19 infections (Ivory, et al., 2021). Still, the lack of international cooperation, of resistance to a comprehensive quarantine on an international basis has allowed for the rise and spread of new variants. The most recent strains after Delta, Omicon variant XEC (CDC, 2021; Katella, 2024) is easily spread and has become the most prevalent in many nations, while an even more virulent mutants, are appearing as KP.2, KP.2.3, KP.3 and KP.3.1.1, Flirt and LB.1.

We should keep in mind the costs of lack of cooperation and consistent quarantine measures in the past. We might review the fact that with Polio, as Burnet (1966) outlined, knowledge of it as a distinct disease did not take place until early in the 19th century, the first epidemic was described in Sweden in 1887 and produced an understanding of its status as an infectious disease. However, it appeared irregularly through a number of countries (e.g. Sydney in 1913-1915) without significant recognition of its potential for damage until the St. Helena epidemic of 1946 and the Arctic epidemic of 1949 where there was high mortality and a 40% paralysis rate was recorded. It was not until the post World War II period that the epidemiology of the different strains of Polio were recognized. One might look to the evolution of these strains as a global experiment for how neglect and ignorance can promote significant costs from nation to nation.

The prospect of Covid-19 persisting for nearly 200 years as Polio appears to have, would be devastating for the global industrial society we live in and require drastic social adaptations to our institutions. Avoiding such a scenario requires significant global cooperation and sharing of resources so that developing nations do not suffer inordinately and do not become reservoirs of new Covid-19 variants.

## A Background of Powers and Institutions: A USA Example

Today many people are concerned about the continuity and change in society and its institutions due to the Covid-19 pandemic and the tremendous change brought about by technology. Among others, Yuval

Noah Harari (2020) has expressed some of these issues in a number of interviews. The nature and variety of societal and individual responses to disease, and how disease affects social adaptions in complex and small-scale societies over time has been a subject of investigation for over one hundred years (Caldararo, 2012).

Malinowski (1945) emphasized the importance of the concept of institutions within the functional analysis of societies, especially during culture change. His definition was quite complex, but describes the relationship people have to each other in both goals and material framework. Elsewhere in the social sciences researchers have had great difficulty in describing institutions and Foley (2001) complained of the difficulty to identify traces of institutions in the fossil record. The same lack of specific agreement is found in discussions concerning theories regarding pre-agricultural societies (Gowdy and Krall, 2016; Houdek, et al., 2016). This uncertainty in prehistory has also been present in a larger scale in discussions of the nature and role of institutions in antiquity and contemporary society. Often this is a consequence of attempts to build theories of state formation and stress on nation-states today and their economic and political stability (Hodgson, 2006).

## America, Power and Oversight

In 1791 the American colonies had just won their independence from the United Kingdom, but faced continued threats from European powers including the French and Spanish. The United Kingdom also was still on friendly terms with many Native nations, including supplying these allies with war material. Attacks by Native people often combined with real or imagined unrest from African slaves in the former colonies (Aptheker, 1993).

During the Revolutionary War leading to American independence, the Native American Allies of the United Kingdom had developed considerable technical and organizational skill in fighting on the British side. After the War was over the inroads of settlers into their territories resulted in considerable loss of life (Hildreth, 1879; McMaster, 1883-1913). The main group was called the Western Confederacy of Native Americans by Europeans and united the Miamis, Shawnees, Delawares and Potawatomis. While the Western Confederacy desired the American colonist Confederacy (not yet formed into the United States of America) to recognize their territory and legitimate ownership and control of western lands, the colonists refused to do so. This was mainly due to the indebtedness of the states of the Confederacy and their need to sell Native lands to speculators and settlers.

The young nation sent an army to suppress the Western Confederacy in 1791, which had successfully driven settlers and some small detachments of militia out or destroyed them. This army under General Arthur St. Clair was defeated in the worst loss in American history with only 24 men escaping without injury. They lost their equipment and camp and suffered the complete rout of the survivors. The Western Confederacy offered peace terms, which were refused, and several later battles took place before the Western Confederacy could be defeated. The defeat of the American army, however, is the topic of interest here and is central to our discussion of institutions and legitimacy (Caldararo, 2019).

The defeat of the army left General St. Clair under President Washington's charge of incompetence. St. Clair blamed suppliers of his troops and officials he felt had mishandled the expedition in general. He desired a trial to clear his name. Has he had been the 15th president of the Continental Congress (Feb 1787 to Nov. 1787), one might imagine that his person could be regarded as having special status. But it is clear that this was an issue that the Congress felt was necessary to deny to all office holders. Washington and Congress came into conflict over the details of the event while Washington forced St. Clair to resign. Much of the governmental structure of the American Confederation was borrowed from England while it had been stripped of general executive powers. Congress desired to investigate the disaster and loss of life and material, but Washington refused. This created the first crisis of powers between the Congress and the President and became the defining point in the creation of the idea of separation of powers in the American government and the concept of Executive Privilege (Rosenberg, 2008). Washington established, in principle, the position that the executive branch should refuse to divulge any papers or materials that the public good required them to keep secret and that, at any rate, they not provide any originals.

Here is the core problem. On what basis can the executive claim such protective powers and what is the "public good"? While a political examination of this conflict is of interest, what concerns us is the fact that the danger of failure of the government was the main reason for refusal. Washington was not just defending his action to send an army in defense of the decision by Congress and the Executive of the United States to seize Native lands. But the legal fabric of the nation had only been in existence for 2 years and the nature of authority was debated but not set by limits of precedent. Rather, the creation of a distinction between the office of the President, and the elected members of Congress representing the semi-independent states, was an act that protected the idea of a unified government and its legitimacy. Also, the first meeting of a cabinet of ministers took place at this time called by Washington and under his direction, although there was no specific provision for it. Washington was acting to create new organs for the presidency.

In the minds of the creators of the American republic that was in formation, the history of state formation in Greece and Rome was an essential reference. In most of these examples, the rule of individual cities could be managed by assemblies, often dominated by one group or another and sometimes called democracies or oligarchies. When alliances were formed between cities, they were usually temporary, yet some were intended to produce lasting peace administered by representatives of cities or groups of cities, the difference between a "symmachy" and a federal state.

Sometimes these were defined by oaths or constitutions, limiting encroachment on the members or binding them in some specific way (Larsen, 1955). Often an executive group to government of a larger unit was formed and given powers to act under certain conditions. These executives, like the Roman consuls could, and often did, use their offices to undermine the independence of the individual cities, or in the case of Rome, to seize state power. The construction of the American Constitution was hoped by the founders to prevent such usurpations (Beard, 1935). Thus the conflict and the contradiction of powers we see today.

## Institutions and Power

Larson's (1955) examples show that the history American founders were attempting to use to balance powers was fraught with failure. The defeat of St. Clair was followed closely by the 1794 Whiskey Rebellion over taxation. There were considerable voices against the use of troops against citizens, yet the survival of the nation seemed threatened by its lack of authority. The creation of an institution, like that of the Office of the Presidency, and prerogatives of power against Congress and the courts was a compromise. The apparatus was dangerous as the examples from antiquity proved. Many feared a strong executive, like Franklin who argued against the Presidential Veto (Beard, 1935).

Today, as in the time of Richard Nixon, the issue of Executive Privilege is foremost in many discussions of the American presidency. And, as in 1791, it cuts to the core of the issue of power, stability and the nature of the institution of the presidency. In 1791, the image of defeat brought about the imagined necessity for a strong executive, but even there had Washington given over the papers requested by Congress it is debatable what, if any, damage might have been done. Yet the power of the executive, in the context of the Covid-19 pandemic, is also enveloped in the nature of information and legitimacy. The Trump in the current case denies any legitimacy of other voices, whether of the knowledge of the Centers for Disease Control, Dr. Fauci or the media at large. The crisis for the institution of Presidential Privilege in the Office of the President rests on the perception of danger to the Republic, now as it did in 1791.

## Conclusion: Crisis and Opportunity

Just as there is a general view that the Black Death (Bubonic Plague) in the 14th century had a dramatic and formative effect on society where some saw the impetus for the opening of the Renaissance (Bowsky, 1971), so today we have an opportunity for change. Our current economy is based on consumer society which is unsustainable, not only in the continuous amount of wealth and labor that goes into the production of goods that often are only used once (as in fast fashion items, see e.g., Schlossbert, 2019) or have built-in obsolescence, but in the waste, pollution and energy demands such a society requires. If, as Keynes argued, we have such a productive capacity that we need not labor for basic needs, it may be time to restructure the goals and demands of a growing population.

Usually when disaster strikes people reorganize along customary lines and rebuild, often with some novel adaptations that have come about over time, as in San Francisco after the 1906 earthquake, or the hurricanes so common along the Gulf Coast. The "hum of the hive" of human social life kicks in and renewal seems a drive to the normal (Caldararo, 2020). But some disasters are so devastating and terrifying, as it seems was the eruption at Santorini or the Black Death, that either abandonment or drastic reordering take place. Still, while civilizations rise and fall due to a number of factors (Lamberg-Karlovsky and Sabloff, 1974; Yoffee and Cowgill, 1988; Tainter, 1988), patterns of regeneration of complex societies argued against complete disorganization and collapse (Schwartz and Nichols, 2006).

Today we are experiencing a vast transformation of how we do work, we are Zooming meetings, training and communication for distribution. Production is so automated that human labor has become less important in so many industries that was inconceivable half a century ago. Through online education our universities could provide new and stimulating learning experiences that could result in new jobs for people to produce virtual products of all kinds, replacing the costly and polluting types as in air travel and auto are still major challenges.

A new kind of consumer society with virtual products has already been in formation in the revolutionary dynamics provided by computers, games and communication. As in the case of the Black Plague, many historians have argued that the foundations of the Renaissance were set in the late Medieval period as mentioned above, the drop in population due to the Plague spurred wage growth, shattered Feudalism and unleashed opportunities for learning and new wealth (Bowsky, 1971). The idea of free time, of leisure often is considered to be morally suspect, but it is a means of opportunity for individuals and groups to discover new ideas and put new organizations into motion.

Our institutions have been under assault in a way parallel to those attacks in the past, as mentioned above in Rome and elsewhere. Instead of a struggle for the old society of the Twentieth Century, it is a time to consider a new society of the Twenty-first Century. Just as government came to the rescue of the financial institutions in 2007, it should now support the medical community and the general public to create "free time" to readjust life, provide for new educational opportunities in our medical research institutions, and a re-evaluation of the goals of society for a constructive future and especially fertility. However, the vast amounts of money spent to salvage the financial institutions has produced only greater inequality. For institutions to function in an environment of greater government and corporate debt there must be new means of subsidizing communities and education to reproduce the resilience necessary for society to rebound after stress (Tainter, 1988). Community-based financial institutions and greater local control can reduce inequality as has been shown in a number of locations, especially in the Spanish Basque areas (Caldararo, 2021).

## References

Acevedo, Monica, Luis Alonso-Palomares, Andres Bustamante, Aldo Gaggero, Fabio Paredes, et al., (2021) "Infectivity and immune escape of the new SARS-CoV-2 variant of interest Lambda," *medRxiv,* July 1, https://www.medrxiv.org/content/10.1101/2021.06.28.21259673v1.

Anon, "A closer look at the public health workerforce shortage," *Keck School of Medicine of USC*, 2020 https://mphdegree.usc.edu/blog/a-closer-look-at-the-public-health-workforce-crisis/.

Aptheker, Herbert (1993) *American Negro Slave Revolts*, New York, International Publishers, 40th Anniversary Edition.

Barrett, Emily Horton, Daniel, Jason Roy, Maria Laura Gennaro, Andrew Brooks, Jay Tischfield, et al., "Prevalence of SARS-CoV-2 infection in previously undiagnosed health care workers in New Jersey, at the onset of the U.S. COVID-19 pandemic," *BMC Infectious Diseases,* 2020 *20, 1:* DOI: 10.1186/s12879-020-05587-2.

Beard, Charles (1935) *An Economic Interpretation of the Constitution of the United States,* New York, Macmillan, 1956 edition.

Berkmeyer, John D., Amber Barnato, Nancy Birkmeyer, Robert Bessler and Jonathan Skinner, "Impact of the Covid-19 pandemic on hospital admissions in the United States," *Heath Affairs,* 2020 v. 39, n.11, https://www.healthaffairs.org/doi/10.1377/hlthaff.2020.00980.

Bloom, Floyd E. and Mark A. Randolph, *Funding Health Sciences Research,* 1990 Washington, D.C., National Academies Press.

Bowsky, William M., *The Black Death: A Turning Point in History?*, European Problem Studies, 1971, New York, Holt, Rinehart and Winston.

Burki, Talha "The online anti-vaccine movement in the age of Covid-19," *The Lancet News,* 2020, 2, 10, E504-5 Oct.

Burnet, (1966) *Natural History of Infectious Disease,* Cambridge, Cambridge University Press.

Caldararo, N. "The hum of the hive," *Anthropology News,* 2020, Sept. 10, https://www.anthropology-news.org/index.php/2020/09/10/the-hum-of-the-hive/.

Caldararo, Niccolo (2019) "The assessment of legitimacy in the 21st century," *Urbanities*, v. 9, Supplement 2, April: 13-24.

Caldararo, N. *The Anthropology of Complex Economic Systems: Inequality, Stability and Cycles of Crisis.* 2014, Lanham: Lexington Books.

Caldararo, Niccolo (2012) *Evolutionary Aspects of Disease Avoidance*, Saarbrucken, Scholars Press.

Caldararo, Niccolo (2021) *The of Leisure and Retirement: Pension Schemes, Community Support, and Contemporary Consequences for the Next Generation,* Washington, Academica Press.

Caldararo, Niccolo (2022*) Origins of White Supremacy and Colonialism, Impact of History's Interpretation on the Present,* Warsaw, Creator Publishing House.

Carline, Jan D. "Funding medical education research opportunities and issues," *Academic Medicine,* 2004, v. 79, n. 10, 918-924.

CDC (Centers for Disease Control) (2021) "SARS-CoV-2 Variant classification definitions," *Science and Research, Variant Surveillance* website, July 13, https://www.cdc.gov/coronavirus/2019-ncov/variants/variant-info.html.

Coppins, McKay (2018) "The man who broke politics," *The Atlantic,* Oct., 17: https://www.theatlantic.com/magazine/archive/2018/11/newt-gingrich-says-youre-welcome/570832/.

Corley, Cheryl "Crime has declined overall during the pandemic, but shootings and killings are up," *NPR*, 2020, July 20, https://www.npr.org/2020/07/20/892418244/crime-has-declined-overall-during-the-pandemic-but-shootings-and-killings-are-up.

Evans, Megan L., Margo Lindauer and Maureen E. Farrell "A Pandemic within a pandemic – intimate partner violence during Covid-19," *NEJM*, 2020 383, 2302-2304.

Foley, R. (2001) "Evolutionary perspectives on the origins of human social institutions," *Proceedings of the British Academy*, 110, 171 – 195.

Garcia, Emma and Elaine Weiss, "Covid-19 and student performance, equity and U.S. education policy," *Economic Policy Institute*, 2020 Sept. 10th, https://www.epi.org/publication/the-consequences-of-the-covid-19-pandemic-for-education-performance-and-equity-in-the-united-states-what-can-we-learn-from-pre-pandemic-research-to-inform-relief-recovery-and-rebuilding/.

Geertz, Clifford (1973) *The Interpretation of Cultures,* New York, Basic Books.

Goldhill, Olivia "'People are going to die': hospitals in half the states are facing a massive staffing shortage as Covid-19 surges," *Stat,* 2020, Nov. 19, https://www.statnews.com/2020/11/19/covid19-hospitals-in-half-the-states-facing-massive-staffing-shortage/.

Grabner, David, *Debt: The First 5,000 Years,* 2011, New York, Melville House.

Halpern, Sue (2021) "Threats against election officials are a threat to democracy," *The New Yorker,* June 29: https://www.newyorker.com/news/news-desk/threats-against-election-officials-are-a-threat-to-democracy.

Harden, John D. and Justin Jouvenal, "Crime rose unevenly when stay-at home orders lifted. The racial disparity is widest in years," *The Washington Post*, 2020, Oct 9, https://www.washingtonpost.com/graphics/2020/local/public-safety/crime-rate-coronavirus/.

Harari, Yuval Noah (2020) "The world after coronavirus," *The Financial Times*, March 20th.

Heale, M.J. (1998) "The politics of patriotism: Loyalty Oaths," in *McCarthy's Americans, American History in Depth,* London, Palgrave.

Hildreth, Richard (1879) T*he History of the United States of America,* Revised Edition, Nabu Press 2012.

Hodgson G. M., What are institutions? *Journal of Economic Issues* 2006, L, 1, 1 – 25.

Houdek, Petr, Julie Novakova, and Dan Stastny, (2016) "Ultrasociality: when institutions make a difference," *Behavioral and Brain Sciences* 39: 24 – 26.

Hsu, Francis L.K. *The Study of Literate Civilizations*, 1969, New York, Holt, Rinehart and Wilson.

Ivory, Danielle, Lauren Leatherby and Robert Gebeloff, "Least vaccinated U.D. counties have something in common: Trump voters," *The New York Times,* April 17, https://www.nytimes.com/interactive/2021/04/17/us/vaccine-hesitancy-politics.html.

Jacobs, Andrew (2021) "They're not feeling 'the summer of joy,'" *The New York Times,* July 6, page D8.

Jorda, Oscar, Martin Kornejew, Moritz Schularick and Alan Taylor, "Zombies at large? Corporate debt overhang and the macroeconomy," *Federal Reserve Bank of New York, Staff Report*, 2020, n. 951, December, https://www.newyorkfed.org/research/staff_reports/sr951.

Kessler, Daniel "Evaluating the medical malpractice system and options for reform," *J. Econ. Perspect.* 2011, 20, 2, 93-110.

Keynes, John Maynard, *Economic Possibilities For Our Grandchildren,* Reprinted in *Essays in Persuasion*, 1930, New York, Norton & Sons, 1963:358-373.

Lamberg-Karlovsky, C.C. and Jeremy A. Sabloff, *The Rise and Fall of Civilizations,* 1974, Menlo Park, Cummings Publishing.

Larsen, J.A.O., *Representative Government in Greek and Roman History,* 1955, Berkeley, University of California Press, second printing edition 1966.

Lintott, Andrew William, *Violence in Republican Rome*, 1999, Oxford, Oxford University Press.

Lopez, German, (2022) "A shift in crime," *New York Times*, Sept 23rd.

Mair, Lucy (1969) *Anthropology and Social Change,* London, The Athlone Press.

Malinowski B., *The Dynamics of Culture Change: An Inquiry into Race Relations in Africa,* 1945, New Haven.

Malinowski, B. (1925) Magic, *Science and Religion*, Boston, Beacon Press.

McMaster, John B. (1883-1913) *A History of the People of the United States From the Revolution to the Civil War*, New York, Sagwan Press reprint 2018.

Meyer, Robinson and Alexis C. Madrigal "The U.S. has passed the hospital breaking point," *The Atlantic,* 2020 December 4, https://www.theatlantic.com/health/archive/2020/12/the-worst-case-scenario-is-happening-hospitals-are-overwhelmed/617301/.

Morstein-Marx, Robert (2004) *Mass Oratory and Political Power in the Late Roman Republic*, Cambridge, Cambridge University Press.

Pardo, Italo and Giuliana B. Prato, *Legitimacy,* (eds) Italo Pardo, and Giuliana B. Prato New York.

Parker, Kim, Rachel Minkin and Jesse Bennett, "Covid-19 continues to hit lower-income Americans hardest," *PEW Research Center*, 2020, Sept 24.

Parsons, T. and E. Shils (1951) *Toward a General Theory of Action,* Cambridge, Mass, Harvard University Press.

Pazzanese, Christina, "Why isn't the right more afraid of Covid-19?," *Harvard Crimson*, 2020 Oct. 30, https://news.harvard.edu/gazette/story/2020/10/what-caused-the-u-s-anti-science-trend/.

Rezba, Claire "Hundreds of healthcare workers have died from Covid-19: This doctor is honoring them," *Healthline*, 2020 Oct. 11, https://www.healthline.com/health-news/hundreds-of-healthcare-workers-have-died-from-covid-19-this-doctor-is-honoring-them.

Rosenberg, Morton (2008) "Claims of Executive Privilege: History, Law, Practice and Recent Developments," *CRS Report for Congress* RL30319.

Schlossberg, Tatiana, "How fast fashion is destroying the planet," *The New York Times,* 2019 Sept 3, https://www.nytimes.com/2019/09/03/books/review/how-fast-fashion-is-destroying-the-planet.html.

Shultz, William J. and M.R. Caine, *Financial Development of the United States,* 1937, New York, Prentice-Hall.

Starr, Paul *The Social Transformation of American Medicine,* 1982 New York, Basic Books.

Schwartz, Glenn M. and John J. Nichols, *After Collapse: The Regeneration of Complex Societies*, 2006 Glenn M. Schwartz and John J. Nichols (ed.). Tucson, University of Arizona Press.

Tainter, Joseph A., *The Collapse of Complex Societies*, 1988 Cambridge, Cambridge University Press.

Thompson, Carolyn (2021) "School board group asks US for help policing threats," *The Associated Press,* reprinted in the *Marin Independent Journal,* Oct. 1, page 6.

Tiftik, Emre, Khadija Mahmood, Raymond Aycock and Sonja Gibbs, (2024) "Navigating the new normal," *IIF Global Debt Monitor* Institute of International Finance, www.iif.com/portals/0/Files/content/Global%20Debt%20Monitor_May 2024_vf.pdf.

Tollefson, Jeff, "How Trump damaged science- and why it may take decades to recover," *Nature,* 2020 Oct. 5, https://www.nature.com/articles/d41586-020-02800-9.

Valerio, Mike, "Excess deaths in DC rise 40%, as residents avoid hospitals during coronavirus pandemic," *WUSA9,* 2020 July 9, https://www.wusa9.com/article/news/health/coronavirus/deaths-from-treatable-diseases-rise-in-dc-fear-of-hospitals-coronavirus/65-caf65ac3-3903-4791-b933-c891f1f62bcb.

Waksman, Ron, (2020) "COVID-19: Leave politics out of practicing medicine," *Cardiovasc Revasc Med.* 2020 Aug; 21(8): 937–938.

Wiggleworth, Robin, Joe Rennison and Robert Armstrong, "Credit crunch hits US small business," *The Financial Times,* 2020 Dec. 15, page 15.

Wood, Stacy and Kevin Schulman, "Beyond Politics: Promoting Covid-19 vaccination in the United States," *NEngJMed*, 2021, 384: e23 DOI: 10.1056/NEJMms2033790.

Yoffee, Norman and George L. Cowgill (eds) *The Collapse of Ancient States and Civilizations,* 1988 Tucson, Un.

# Chapter Eight

# Climate Change, Economic Philosophy and Policy

As the chapter on the Khmer demonstrates, people do attempt to change their behavior as environments change. We lack the ethnohistorical documents that might explain how the Khmer came to make investments to avoid collapse, but one can assume that, as in the case of the people of Teotihuaco (Millon, 1988) environmental stress, even if largely human driven, can be mitigated to some degree if resources can be diverted and reward goal systems can be modified sufficiently. This is not entirely the point, however, as long term adaptations can sustain a society even if the population remains unstable or its growing demands on resources can be met by more efficient exploitation or by increasing external sourcing by trade or war. Both the Sumerian and Teotihuacan examples provide us with models of this given the long existence of both entities. In the case of the Sumerians the systematic cycles of intragroup war, as in the case of the Mayan kingdoms, undermined such resource adaptations (McC. Adams, 1988; Yoffee, 1988; Schele and Freidel, 1990; Culbert, 1988).

Biesbroek, et al., (2013) attempt to enumerate the varied factors involved in blocking or complicating climate change adaptation. Vink, et al. (2013) describe the lack of comprehensive approach in the literature concerning knowledge and power to implement initiatives. A good example of this situation, regarding both policy and power, was the conflict between the United States Federal Reserve's Jay Powell rejecting any responsibility of the Fed to respond proactively to climate change initiatives and that of Isabel Schnabel, member of the European Central Bank's six person executive board who advocated greater action and involvement in January of 2023 (Giles and Mosolova, 2023).

The recent wild fires in California, including that in Paradise and the very destructive fires in Portugal, Russia, Brazil, Greece and Indonesia have illustrated again the human toll and cost of wildfire. This chapter is not about Paradise, rather it is about because of existing policies and fire dogma, economic philosophies and policy, there will be more Paradises to come. Until there is a comprehensive reanalysis of these policies and dogma, which this paper is a brief outline, the crisis will continue and cannot be attributed to only climate change which is a distraction from the main causes. However, most discussions today in the media on the subject of forest or wild fire ignore the forest/development industry's history concerning forest and wild fires as influenced by forest management, agriculture and commercial uses of open space. It ignores the dramatic changes that industry has brought to our grasslands and forests over the past 200 years.

Much of the data used to justify logging, defensive space and prescribed burns is based on an unscientific methodology called "fire scars" which identify tissue trauma in trees as fire evidence. This concept is based on a subjective interpretation. Such trauma evidence can be caused by limbs striking trunks, disease, etc. Studies of ancient burns are not verified by carbon analysis as described by scientists such as Goldberg (1985). Fire (as industrial processes and recreational uses) as well as burning of fossil fuels and wildfires all contribute to warming of the Earth and climate change. The increase of dust in the atmosphere also can contribute to this, mask the effects of warming and slightly reduce it (Kok, et al. 2023). As a quantity resulting from land clearance, industry and erosion in general, dust has been difficult to chart as an element in the past cycles of warming (Abadi, et al., 2019).

A recent study (Stephens, Martin, & Clinton, 2007) argued that 1.8 million hectares burned annually in prehistoric California, a tremendous amount that would have made difficult the survival of wildlife to support Native Americans and the ability of new biomass to survive to feed wildlife until new growth could take place (see General Surveys in Heizer & Whipple, 1971). Other commercial changes were engineered by the advance of agriculture from the Colonial period. Our private forest lands been routinely logged over and then had the debris burned off for over one hundred years, thus continuing the process of creating a fire adapted ecology begun with the settlement of the country. But also, our public forests, both national and state have been subject to commercial logging since their creation with few exceptions. Added to this has been a long history of spraying pesticides onto forests to kill insects that attack trees and herbicides to kill unwanted trees and brush. S.H. Dreistadt, et. al., documented the extent of this in a 1990 study. Herve Jactel (2000); Peter Gorrie (1993) and Peter Montague (1999) have summarized why pesticides are counter productive. The issue is not that there is a history of fire, rather that logging and fire are part of a cultural pattern that has created a problem.

## Fire Suppression and Prescribed Burns

Another part of this dogma is the idea that the forest industry or the Forest Service has tried to keep fire out of our forests and that this is now responsible for the forest fires we now see. Paul Mellars in a 1976 article detailed how our forests have been fired yearly to increase the growth of forage for commercial herds and wild populations of game for hunters for over 100 years. Thus the idea that prescribed burns are new is simply false, and J.E. Keeley, et al., (1999) found in a study of prescribed burns that they did not always reduce fire danger and often increased not only the potential but the intensity of wild fires. Riggen, et al., (1988) found in their studies of prescribed burns that there were significant variations in their effects, just one of which was seen when they burn out of control as in the Cerro Grande Prescribed Fire (NPS, 2001).

One would expect that if natural fire regimes were working they would burn off all the fire fuel and leave areas without return fire. Alternatively, if areas have not burned within established fire regime periods related to fire fuel build up then they would be expected to burn. Minnich, et al. (2000) competed a very thorough analysis of photographic images taken from the air over the past 60 years in the southern desert areas of California. Their maps indicate that neither expected result can be supported. Some areas never burned and others burned several times over the period 1925 to 1990. While few fires were noted between 1955 and 1971, the previous periods from 1925 to 1955 were characterized by many small and large fires. The period of the 1990s which is expected to have attained maximum fire fuel levels from the 1960s fire suppression practice shows fewer overall fires but overlapping fires with periods 1971 to 1985. While Minnich, et al. (2000) argue that overlap fire areas did not result in considerable burned areas, the fact that they did occur would seem to be contradictory to the general theory. Similar evidence is available from the Southwest in the La Mesa publication (1996) and from recent Southern California fires especially in the San Diego area where fires have revisited the same areas with devastating results. It is clear that in many locations in the world there is good evidence that wild fires spread unpredictably and that the factors involved are complex for any single fire event (Crawford & Pianka, 1996; Agee, 1996). Analysis of examples of fire history discovery techniques (e.g., pollen studies, carbon in soil, etc.) are found in Caldararo, 2002 and 2004. Variations in fire regimes appear to be considerable when relying on "fire scars" in contrast to expected climate driven drying and/or fuel buildup (Brown, 2002) and FHX2 software does not seem to produce explanatory evidence. This is due to the fact that the central feature used, the "fire scar" or "fire char" (Swetnam, 1993) is not a scientifically defined or described phenomenon (Lachmund, 1921; Baisan & Swetnam, 1990). The same feature can result from any trauma or infection that can bring on necrosis of tree tissue (Blanchard & Tattar, 1981; Sutherland, 2016)).

The ethnohistorical data is also biased where Native American memories are affected by histories of contact and no longer "pristine records of past events." This can explain why studies that rely on fire scars and ethnohistorical data (e.g., Stephens, et al., 2007) often produce extreme fire histories where up to 88% of California is annually burning. Though more difficult to collect and assess (Goldberg, 1985), studies of fossil pollen and carbon in the soil produce less biased results though are still plagued with the problem of agreement (Caldararo, 2004; Condera, et al., 2009).

The S.F. Chronicle (e.g. Fimrite, 2007; Martin, 2007) and other newsmedia (e.g., New York Times, Johnson, 2006; Associated Press, Flaccus, 2006) in their coverage of these wild fires in the State of California, and interviews with the victims who lost homes, focus attention on another element of the dogma with the idea of clearing brush around structures to produce what is called "defensible space," and criticism of environmental pressure to limit logging. In both cases there is no scientific support for these practices. Fires do not respect "defensible spaces," as recent fire disasters have proven and yet promoting such practices gives people a false sense of security and may even lead to people believing they can ignore warnings to evacuate. Though staying in a house and saturating the structure and surrounding area does seem to be associated with the survival of some homes, the evidence is not available to provide statistical support. However, a new movement (Shelter-in-Place) has grown up to avoid both evacuation and loss of homes by claiming certain designs of housing can resist any fire or other disaster (Weisberg, Showley & Pierce, 2007).

Fires do not burn one bush or tree at a time, they consume both organic materials in their path and available oxygen being also driven by wind and contours of the terrain. Studies published from the second La Mesa Fire Symposium by the United States Forest Service in 1994 detail these factors (Craig, 1996). To believe that a few hundred feet of cleared space (California law changed in 2005 required 100 feet of "fuel clearing" and 30 feet of a "clean and lean zone" with little flammable material present) will save a house is ridiculous in this context. The presence of such space may be a factor in the choice of fire fighters to stay and try and save a house and may lead to loss of life among them.

The present argument that the Forest Service has created disasters by years of fire suppression is also unfounded. The practice the Forest Service has followed is that of forest management established since the virgin forests of America were burned and logged off by settlers in the 18$^{th}$ and 19$^{th}$ centuries, as mentioned above. That practice was to fire the forest undergrowth to promote forage for wildlife that supported hunting, both commercial and recreational. The greatest increase in biomass occurs after both burning and logging (Longhurst, 1961). Human land conversions greatly increase fire fuel production/consumption cycles and logging is one important process in preconditioning forests for high-intensity fire (Stocks and Kauffman, 1997).

The Forest Service has invested huge sums in fighting fires but it is questionable if these efforts contain fires more than natural burning limits would produce, given the samples of fire histories produced in recent years. Pyne (1982, 1991,&1995) has suggested this same conclusion from his research in terms of cost of fire fighting and the results. Rather the facts are that we have created fire adapted forests by commercial uses of the forests, prescribed burns and by a history of logging. In 1996 *New York Times* reporter James Brooke published a story arguing that the "total suppression policy of the 1960s has led to big buildup of fuel." Included in the article was a chart from the National Interagency Fire Center. From it he argued it showed that fire frequency was most "suppressed" or burned fewer acres in the 1960s, but that fire frequency in the 1970s was no lower than in the 1990s but lower than the 1940s and 1950. Reference to the chart, however, demonstrated that the greatest amount of acreage burned was in the 1950's and the lowest in the 1980's and has again risen after a decade of "let it burn" philosophy and prescribed burns became policy (Caldararo, 2004).

Mike Davis (1995) summarized the fire frequency data for much of California finding that in the Malibu area, the rugged 22 miles coastline "is scourged, on average, by a large (1,000 + acre) fire every two years, and the entire western Santa Monica Mountains has been burnt three-times over this century." He lists the fires and dates with acres affected. James Sweeney noted a similar fire history of reburning areas in his 1956 summary. We must ask, then, what fire suppression? And what is the relation of fire suppression to the frequency of fire?

The fire adapted nature of our forests and wildlands is the product of 50,000 years of human treatment of the natural environment. I documented this history in a paper on the human ecology of fire (Caldararo, 2002) and its cultural origins (Caldararo, 2004). The fire adapted nature of the forests and the lack of prevention by prescribed burns was demonstrated in the recent Los Alamos fires where a prescribed burn turned into a major forest fire. The same is true of the Grayline Fire that also started from a controlled burn (see Stephan Bridge Fire Study). Logging has been found to be one of the factors associated with the highest frequency of fire (Stocks & Kauffman, 1997). This is only logical as once logged, replanted areas simply produce a tree farm with all trees of the same age and grow to the same height supporting "crown fires" of closely spaced trees and interlocked upper stories. Natural forests do not grow this way, and climax forests, often called "old growth" forests are characterized by a distribution of old and young trees. In both cases Marston Bates in his *The Forest and the Sea*, and Peter Mattheissen in his *Wildlife in America*, described these relationships. The build up of fire fuel and the question of fire prevention, as opposed to natural regimes, has been studied by Johnson, Fryer and Heathcott (1990) in British Columbia. Their study found that in contradiction to the idea that fire suppression was responsible for fire frequency and intensity, they found the fire history to be a natural regime.

World-wide, nearly 80% of all fires are the result of human behavior either by directly setting a blaze as arson, or by accident (Caldararo, 2004). In California, the California Department of Forestry and Fire Protection released figures in July 2007 showing that 13.2% of all fires in the State were arson, 29.1% from equipment use; campfires, smoking, playing with fire by adults and children accounted for 5.6%, debris burning 9%, vehicles caused 11.8%, miscellaneous human causes at 12.2% and undetermined at 16.9%. Essentially we are fire bugs. Our activities cause fires and as we build more housing into the rural areas we are the cause of more "wildfires." We should be investing funds to find ways of deterring this propensity, to use education and the media to condition people to avoid use of fire in all but productive enterprises.

For years environmentalists have been trying to get people to live with the land, to reduce the impact of roofs and energy use. Surrounding a home with greenery, especially using trees accomplishes this result. The "defensible space" concept undermines these efforts.

In post-fire studies of surviving vs burned structures from the 1961 Bel-Air fire and the 1990 Santa Barbara fire three factors affected survival, the most significant was a nonflammable roof, the second a defensible space and third someone home to fight the fire. The last factor accounted for 9% but the nonflammable roof vs defensible space were not separated as components, though a nonflammable roof was cited as the main factor associated with the pathway of embers causing fire.

When the Cleveland National Forest fire in 2003 in San Diego county was finally put out, over 2,800 homes had been destroyed. There was no statistical difference between those homes that had "defensible space" and those that did not. Data from one of the best studied fires, the Black Tiger Fire published by the National Fire Protection Association, produced tables of survivor qualities for houses in the fire. It was a human caused fire in 1989, covering 2100 acres and destroying 44 homes. Of those houses that survived fire the associated factors were slope characteristics and exterior construction of non-flammable construction. The Stephan Bridge Road Fire burned more rapidly than predicted due to lower fuel loads which points out the illusion that fire fuel loads alone are the single determining factor in fire behavior. We are denuding the landscape in fear of the fire that we are creating ourselves at a time when decreasing municipal and county budgets make available fewer resources and less funding for fire fighting. Clearing space is not the answer, limiting building outside of the urban core is one answer, increasing penalties for causing fires is another. Fines and jail time would result in less spark producing equipment and more careful operation. The boys who started the Mt. Vision fires in Marin were not prosecuted and their parents were not charged for the fire they set. It would seem that policy professionals could intervene to end this tolerance of arson with a variety of methods.

## Housing and Developments in Wildlands

A related problem Martin (2007) brings up is low cost housing in the Sierra and other forested and wildland areas. The Sierra counties and cities had the opportunity to employ zoning restrictions and impact fees to provide for low cost housing decades ago. The tradition of anti-government property fanaticism defeated any potential for this and was supported by the poorest and most in need of such housing. The local newspapers attacked all proposals for regional planning to provide affordable housing and limit sprawl that now makes the protection of property in fires so impossible. Greater control of housing developments in wildland and forest areas is necessary to limit costs of protecting lives and property and the lives of fire fighters.

## Bias and Random Factors in Data on House Survival

I think the most significant handicap the public is faced with is the belief (or propensity to imply when speaking to the popular press) among forest science advocates that fires burn one tree or bush at a time. What we see instead, especially from temperature, wind and aerial analysis is a "storm." A fire storm proceeds along the contours of the land and is deflected or drawn by a number of factors. Frederick J. Swanson (1981) described this in great detail. Biomass consumption takes place based on the nature of the fuel that the fire storm has consumed and is consuming but also on its directional velocity (see Stocks and Kauffman, 1997.) The nature of fires thus varies considerably one from another and has been found to be based on the unique climatic, geomorphic and chemical components of the locale. These features result in fractal plasma streams that can be reproduced in the laboratory (Gaier, LeinStockman, et al., 2004). In using data from forest fires and applying it to other systems, Kan Chen and Per Bak (2002) found that characteristics of forest fires show a novel scaling form for the distribution of dissipation in the forest model. These are turbulent systems of energy that are dissipated (burn) locally. This is why forest fires damage homes in a variety of conditions and why Jack Cohen's tests (a research physical scientist at U.S. Forest Service Fire Sciences Laboratory in Missoula Montana, Cohen, 1999; 2001) showed that 10 to 15 meters should be sufficient under some conditions and this is used by many proponents of defensible space as a measure. However, Cohen's research is not published, thus not peer reviewed and not available from the USDA for examination as it is not listed on the USDA index nor the Fire Service abstracts, though related research is summarized by Alexander (2001). Nevertheless, we might assume Cohen has used some standard methods by the USDA scientists (and UC's Forest Products Laboratory) to expose brush and wood to heat and monitor flame generation and kindling point. This is not an effective measure of how a fire storm "dissipates" energy as burning. Heat, duration of exposure and energy transfer are all different in a fire storm than simple exposure of combustibles to heat. This is why the outcomes of actual fires differ so much and show, even within one particular fire, significant variation. For example:

1. The 1961 Belair-Brentwood fire is said to have had homes survive with non-flammable roofs and a clearance of 10- to 18 meters (30 to 60 ft.) (95% survival rate, Howard, et al, 1973).

2. The 1990 Painted Cave Fire found homes with wood roofs had a 20% survival rate, and those with non-wood materials a 70% survival rate, combined with a vegetation clearance of 10 meters or more the survival rate went up to 90%. Still homes with flammable roofs, no clearance and no defense still survived at a rate of 4% (Kirsch, n.d.).

3. The Black Tiger Fire, one of the best studied fires where fireline intensity (FI) was measured found that where FI was at least 500 Btu/feet per second 11.7% of the structures survived. Of these one had a clearance of 10 feet the other 75 feet. Both had non-flammable roofing materials, and were on slopes of 10 to 15%, considered good for survival. With an FI of less than 500 Btu/fs, 18% of the homes survived one with 5 feet of clearance and one with 10 feet. Of the homes that survived 25% had less than 10 feet of clearance and of those with none more than 75%. Average clearance was 25 feet with 75% under 10 feet. At that FI, however, roof construction materials are not significant either as only 16% survived with non-flammable roofs.

What is needed is to define safety, but why also we see homes burn at 50 feet and not at 5 feet. Into this equation we must also figure fire fighter behavior.

In a November 2007 OpEd piece in the *Marin Independent Journal* Scott Alber and Kent Julin, representatives of Marin County's open space and fire management agencies, answered my earlier discussion in the same paper of the flaws in the concept of "defensible space." This is an issue that requires more public discussion. It was also gratifying to find that they did not refute any of the scientific evidence or studies I used to criticize it. The massive destruction of trees and plant life that results from the creation of "defensible space," is unnecessary. Alber and Julin say that "study after study has shown that the solution to disastrous home loss during wildfires is to prevent buildings from igniting..." No one would contest this conclusion or such studies. Obviously fires burn combustible materials.

The main flaw in the defensible space idea is that its proponents claim that clearing space around a home protects it from wildfire. This is simply not supported by the scientific evidence. As I cited the Black Tiger Fire (though others show the same result), some homes with cleared areas burned and some with no cleared areas (less than 10 feet) did not. Other factors like combustible roofing materials, slope, decks, open eaves, are more significant.

Since Alber and Julin bring up the Oakland Hills fire in 1991, reference to its causes should be examined. First, the fire was started by workers who imprudently operated machinery. Had they followed proper procedures no fire would have occurred. Second, the fire department response team did not put the fire out properly and left before embers were totally dead. Third, when residents called in the revived fire their calls were ignored because the fire department believed they had already put it out. By the time they responded it was out of control. The fire was the result of negligence all round not because people had not cleared trees and brush from the land. If we only denude the land of greenery we will not stop wildfires, but if we stop setting fires and we fight them properly and stop building homes in the wildlands and forests, we will minimize the problem.

Following this exchange, Mr. Tim Thompson, battalion chief with the Marin County Fire Dept. made a remarkable admission in the November 11th 2007 the *Marin Independent Journal* newspaper about the decisions made on the ground by firefighters. In the face of concerns of property owners about which homes were chosen to defend and those allowed to burn his statement is revealing. He asserts that the primary factor he and his colleagues used to determine which houses to defend was defensive space. One would expect that it is prudent to use all the existing information about which homes survive fires to make such a decision. It is clear from the data on fires in the past 10 years that defensive space is not a major factor. Location on slope is associated with the best survival followed by roof materials, siding, decks, eaves and terrain. Property owners should be able to be confident that their properties are being given equal protection by fire authorities. If Mr. Thompson and his other battalion chiefs are making decisions primarily on defensive space they are skewing the data on survival and influencing outcome as well as following a self-fulfilling prophesy strategy of fire protection practice. This certainly is unexpected from a professional, we would imagine that they would arm themselves with the best available information to execute their jobs, not act ideologically. This tendency may put firefighters more at risk and may result in homeowner deaths if they think that defensive space will protect them no matter what. What Mr. Thompson has done is place the State of California in a tenuous position in upcoming liability issues over the recent fires. One would hope that he and his associates will reevaluate their positions and act in the future with less bias and in response to more evidence. One would expect that the substantial resources applied to restudy fire response by at least the Canadian Forest Service (Alexander & Thomas, 2007), especially in social science research in evaluation of practice and effectiveness of response would nullify such bias and be appreciated by American fire service agencies.

A recent review of fires in California by Syphard and Keeley (2019) came to the conclusion about defensible space:

> "The relative importance of defensible space, as quantified by deviance explained in the regression models, was virtually nil statewide...,"

In a later restudy (Syphard, et al., 2021) they concluded,

> "Generally, the pattern of housing relative to vegetation better explained structure loss than local-scale vegetation amount, but the results varied regionally."

Scarce resources that the public has available for fighting fires should be efficiently utilized, and while we should defer to the expertise of the professionals it is appalling to find their decision-making based on a narrow view. Espousing the defensive space philosophy also simply encourages building in the most fire-vulnerable landscapes and discourages the use of biomass as a means of reducing the impact of existing human built space, providing some coverage for wildlife as well as reducing energy concentrations from roofs and structures that plays a role in global warming.

## Vicious Cycle: Logging, fire-adapted ecosystems and Fire-Fighting

Reporter Mark Prado (12/20/07) reported on the discovery that a fungus causing brown cubicle heart rot was attacking the redwoods in Muir Woods, the implication was that a lack of fire was the culprit. While this is a disturbing finding, the response is both typical and wrongheaded. The idea that fire is the answer to all the forestry and wildland problems we have has become dogma. It is evident that human activities have upset the natural world, but there is no archaeological evidence to prove that fire was more frequent in the past than in the past 400 years of European settlement and there is evidence that it was less frequent prior to the arrival of Native Americans. I produced a critique of this argument and assessed the sources in my 2004 book. Other arguments, like that asserting forests without regular fire are unhealthy systems have no more support than other theories that may better explain the same conditions. For example, Forest Service and commercial forestry use of pesticides may be to blame. Paul Fine and Pyhllis Coley found that herbivores (including insects) control types of forest growth supporting diversity (Fine & Coley, 2004).

From an anthropological perspective this is a rather primitive way of going about to solve problems. The attitude that we must fight this fungus with fire, like the approach that we must muster our forces against wildfires as if we were fighting an enemy is part of a long cultural tradition that we must tame nature. The enemy is out there, and we need to circle the wagons and spray it with pesticides or bomb it with fire retardant, bulldoze fire breaks, and deny it fuel as a defensive measure like armies used to scorch the earth to deny an invader food. Instead we should realize that we are the cause of these problems. If we did not build homes in the forests and wildlands we would have fewer problems. For nearly one hundred years the forest service, open space and commercial forest products companies have sprayed these lands with pesticides, killing off natural predators of insects and pathogens that affect forest health. They have killed off the game that ate the undergrowth that fuels fire and they have fired adapted our lands by continuous burning both after logging and to produce more game.

In this instance regarding brown cubicle heart rot, the scientific literature tells us that this fungus is spread by human activity as well as Sudden Oak Death Disease (http://ipm.ucanr.edu/PMG/PESTNOTES/pn 74151.html). Professor Garbelotto at UC Berkeley in 2004 wrote that the fungus is propagated on the stumps of newly cut down trees and transmitted by infection to the roots of nearby trees. Therefore, across our State, and perhaps the country, loggers are spreading the disease. Further, this may be a product of the policy of thinning forests now part of the Forest Service since the Bush Administration. If we want to stop this disease, and return our forests and wild lands to some degree of stability, we need to reduce the level of human activity in them. This means severely reducing public access to Muir Woods and other parks and open space and reducing the level of logging statewide and placing a moratorium on building outside of our current urban areas. Perhaps an analysis of logging equipment is in order, or rigorous cleaning of logging contractor machinery as they move from one location to another. We are the problem, let's stop the war on nature and start thinking about the consequences of what we are doing.

# References for Part I

Agee, James, "Fire regimes and approaches for determining fire history," *General Technical Report, INT-GTR-341,* 1996 (revised 2002).

Allen, Craig D., editor, *Fire Effects in Southwestern Forests, Proceedings of the Second La Mesa Fire Symposium*, Los Alamos, New Mexico, March 29-31, 1994 USDA Forest Service, Ft. Collins, Colorado, Sept, 1996.

Alexander, Al, "Fire safe structures: Proposed changes to Building codes in California," *University of California, Forest Products Laboratory*, 3 Nov. 2003.

Alexander, M.E. & Thomas, D.A., "Wildland fire behavior, case studies and analyses: value, approaches, and practical uses," *Fire Management Today*, Natural Resources Canada, Canadian Forest Service, 2007.

Baisan, C.H. & Swetnam, T.W., "Fire history on a desert mountain range: Rincon Mountain Wilderness, Arizona, USA," *Can. J. For. Res.,* v. 20, n. 10, 1990:1559-1569.

Blanchard, R.O., & Tattar, R.A., *Field and Laboratory Guide to Tree Pathology*, Academic Press, N.Y., 1981.

Bates, Martson, *The Forest and the Sea*, Random House, New York, 1960.

Biesbroek, G. Robbert, Judith E.M. Klostermann, Catrien J. A. M. Termeer & Pavel Kabat, (2013) "On the nature of barriers to climate change adaptation," *Regional Environmental Change*, v. 13, 1119-1129.

Brooke, James, (1996) "Western wildfires near record season," *New York Times,* Oct 23.

Caldararo, Niccolo, (2004) *Sustainability, Human Ecology and the Collapse of Complex Societies, Economic Anthropology and a 20th Century Adaptation,* Mellen Studies in Anthropology, v. 15, The Edwin Mellen Press, Lewiston.

Caldararo, Niccolo, "Human ecological intervention and the role of forest fires in human ecology." *The Science of the Total Environment*, v. 292, 2002: 141-165.

Cohen, Jack D., "Reducing the wildland fire threat to homes: where and how much," *USDA Forest Service General Tech. Rep.,* 1999.

Cohen, Jack D., "Wildland-urban Fire Research: Determining Home Ignition Potential During Wildland Fires," *US Forest Service Fire Sciences Lab.* 3, Nov. 2003.

Crawford, Melba & Pianka, Eric, "Great Victoria Desert, Australia: Topic, Wildfire Behavior," CSR/TSGC,www.csr.utexas.edu/projects/rs/aussie.html, July 19, 1996, 3 pages.

Culbert, T. Patrick, (1988) "The collapse of Classic Maya civilization," in *The Collapse of Ancient States and Civilizations,* Norman Yoffee and George L. Cowgill, (eds) Tucson, The University of Arizona Press: 69-101.

Davis, Mike, (1995) "The case for letting Malibu burn," *Environmental History Review*, v.19, n. 2,1-36.

Dreistadt, Steve H., Dahlsten, Donald L., & Framkie, Gordon W., "Urban forests and insect ecology," *Bioscience,* Mar. v. 40, 1990:192-9.

Fimrite, Peter, "Forest fires bigger, hotter, study finds," *San Francisco Chronicle,* August 9th 2007.

Fine, Paul & Coley, Phyllis, "Herbivores promote habitat specialization by trees in Amazonian forests," *Science,* v. 305, n. 5684, 2004: 663-665.

Flaccus, Gillian, "Wildfire likely triggered by controlled burn," *San Francisco Chronicle*, February 8th 2006.

Garbelotto, M., (2004) "Root and butt rot diseases," Forest Pathology, *Encyclopedia of Forest Sciences,* 750-758.

Giles, Chris and Daria Mosolova, (2023) "Fed and ECB at odds over their roles in tackling climate change," *The Financial Times*, January 16, page 3.

Goldberg, Edward D., *Black Carbon in the Environment,* John Wiley & Sons, N.Y., 1985.

Gorrie, Peter, "Is spruce budworm the real enemy?" *Canadian Geographic,* Mar/April, v. 113, issue #2, 1993: 6-8.

Howard, Ronald, North, U. Offensend, Fred L., et al., *Decision analysis of fire protection strategy for the Santa Monica Mountains, an Initial Assessment*, Menlo Park, Ca., Stanford Research Institute, 1973.

Jactel, Herve, "Trees and Forests: Living with insects and microorganisms," IUFRO World Series, v. 11, *Protection of World Forests from Insect Pests: Advances in Research*, #3, 9-12.

Johnson, E.A., Fryer, G.I. and Heathcott, M.J., "The influence of man and climate on frequency of fire in the interior wet belt forest, British Columbia," *Journal of Ecology*, v. 78, 1990:403-412.

Keeley, J.E., Fotheringham, C.J. and Morais, M., "Reexamining fire suppression impacts on brushland fire regimes," *Science,* 284, 1999:1829-1832.

Kan Chen and Per Bak ("Forest fires and the structure of the universe," *Physica A: Statistical Mechanics and its Applications*, V. 306, 1 April, 2002:15-24.

Kirsch, Susan, "Preventing Wildfire Home Ignition: Why the Answer is 'over your head,'" *Anchorage Fire Dept.,* Anchorage, Alaska, n.d.

Kok, Jasper F., Trude Storelvmo, Vlassis A Karydis, Adeyemi A. Adebiyi, Natalie L Mahowald, et al., (2023) "Mineral dust aerosol impacts on global climate and climate change," *Nature Reviews, Earth & Environment,* v. 4, 71-86.

Gaier, LeinStockman, et al., "Ultrafast mulitphoton forest firess and fractal in clusters and dielectrics" *J. Physi. B. At. Mol. Opt. Phys,* v. 37, 2004, L57-67.

Lachmund, H.G., "Some phases in the formation of fire scars," *J. of Forestry*, v. 19, 1921:155-168.

Longhurst, W.M., (1961) "Big game and rodent relationships to forests and grasslands in North America," *La Terre et la Vie,* 305-319.

McC. Adams, Robert (1988) "Contexts of Civilizational Collapse: A Mesopotamian View," in *The Collapse of Ancient States and Civilizations,* Norman Yoffee and George L. Cowgill, (eds) Tucson, The University of Arizona Press: 20-43.

Martin, Glen, *San Francisco Chronicle Magazine,* 29 July 2007.

Matthiessen, Peter, *Wildlife in America*, The Viking Press, New York, 1959.

Mellars, Paul, (1976) "Fire ecology, animal populations and man: a study of some ecological relationships in prehistory," in *Proceedings of the Prehistoric Society*, eds., John Coles and John G. Evans, v. 42, December,15-45.

Millon, Rene (1988) "The last years of Teotihuancan dominance," in *The Collapse of Ancient States and Civilizations,* Norman Yoffee and George L. Cowgill, (eds) Tucson, The University of Arizona Press: 102-164.

National Fire Protection Association, *Black Tiger Fire Case Study,* Fire Investigatons Division, National Fire Protection Association, Quincy, Massachusetts, 1992.

National Fire Protection Association, *Stephan Bridge Road Fire: A Case Study*, National Fire Protection Association, Quincy, Massachusetts, Posted 2007.

National Park Service, *Cerro Grande Prescribed Fire*, Board of Inquiry Final Report, February, 26,2001.

Olshansky, S. Jay, Bradley J. Wilcox, L. Demetrius and Hiram Beltran-Sanchez, (2024) "Implausibility of radical life extension in humans in the 21st century," *Nature Aging,* 7 Oct., doi.org/10.1038/s43587-024-00702-3.

Olshansky, S. Jay, Bruce A. Carnes and Christine K. Cassel (1993) *The aging of the human species, *Scientific American,* April, 46-52.

Riggen, P., Goode, S., Jacks, P., Lockwood, R., (1988) "Interaction of fire and community development in chaparral of southern California," *Ecological Monographs*, v. 58,155-176.

Schele, Linda and David Freidel, (1990) *A Forest of Kings, the Untold Story of the Ancient Maya,* New York, William Morrow & co.

Stephens, Scott L., Martin, Robert E., & Clinton, Nicholas E., "Prehistoric fire area and emissions from California's forests, woodlands, shrublands and grasslands," *Forest Ecology and Management*, doi:10.1016/j.foeco, 2007.06.005.

Stocks, B. and Kauffman,J.B., "Biomass consumption and behavior of wildland fires in boreal, temperate, and tropical ecosystems: perameters necessary to interpret historic fire regimes and future fire scenarios," in J.S. Clark, H. Cachier, J.G. Goldammer, B.Stocks, editors, *Sediment Records of Biomass Burning and Global Change,* NATO ASI Series, v. 151, Berlin: Swpringer-Verlag, 1997:169-188.

Swanson, Frederick J. "Fire and geomorphic processes" in *Proceedings, Fire Regimes and Ecosystems Conference,* 1979 Dec. 11-15 Honolulu, USDA June 1981.

Swetnam, T.W., (, 1993) "Fire history and climate change in giant Sequoia groves," *Science,* v. 262:885-889.

Vink, Martinus J., Art Dewulf and Catrien Termeer (2013) "The role of knowledge and power in climate change adaptation governance: a systematic literature review," Ecology and Society, Dec. v. 18, n. 4, https://www.jstor.org/stable/26269416.

Weisberg, Lori, Showley, Roger & Pierce, Emmet, (2007) "Exclusive homes emerge unscathed as fire protection concept is tested," *Union-Tribune,* San Diego, Ca., Oct, 25.

Yoffee, Norman, (1988) "The collapse of Ancient Mesopotamian states and civilization," In *The Collapse of Ancient States and Civilizations,* Norman Yoffee and George L. Cowgill, (eds) Tucson, The University of Arizona Press: 44-68.

# Part Two:

# Wildfire and Fire-adapted Ecology: How People Created the Current Fire Disasters

## Introduction

The Waldo Canyon fire in the foothills of Colorado Springs, Colorado in 2012 is a prime example of the nature of wildfire today (Randall, 2014). Viewing the now iconic image of the fire destroying homes and cars across the suburban development gives a perfect picture of the problem. Most people think of wildfire taking place in forests or in rural areas where it is a natural environment that burns. The problem here is that there is no natural environment any more, humans are everywhere and all land is used for some human agency. The result of this human activity is a disturbed ecology and central to this is a fire adapted regime.

**Figure 7: Waldo Canyon Fire June 26, 2012**

## Methods

The methods used in this part of this paper included walking the Mount Vision fire area in Marin County, California as well as the recent Valley Fire surrounding Middletown, analyzing reports of fire agencies of the fire response, interviews with fire fighters and home owners. Some soil samples were taken to verify reports in the case of the Vision fire. Comparing data on fire frequency and intensity from various agencies was utilized as a means of understanding historic trends in fire burns in areas and build up of fire fuel in biomass.

## The Problem

In the mid 1960s I began a program of research into forest management after having visited a number of lumber camps and lumber mills in the Humboldt County area of California. The camps were squalid affairs where living conditions for workers and their families seemed to be a holdover from the 1930s. The mills were just as disturbing, not just the constant noise, but the ever present danger of serious injury and death. About this time the lumber mill in the Ferndale area had an explosion and spectacular fire. I viewed this fire with friends who had worked in the mill and it was not an unusual event. These are still common events due to a buildup of sawdust, equipment failure and poor safety procedures. Images and footage of recent fires and explosions at mills are available online (https://video.search.yahoo.com/search/video;_ylt=AwrBT8Ib9tVVpDkAdZZXNyoA;_ylu=X3oDMTEyODRjcmJxBGNvbG8DYmYxBHBvcwMxBHZ0aWQDQjAzNDFfMQRzZWMDc2M-?p=Explosion+and+Fire+at+the+Simpson+Mill&fr=yfp-t-901-s).

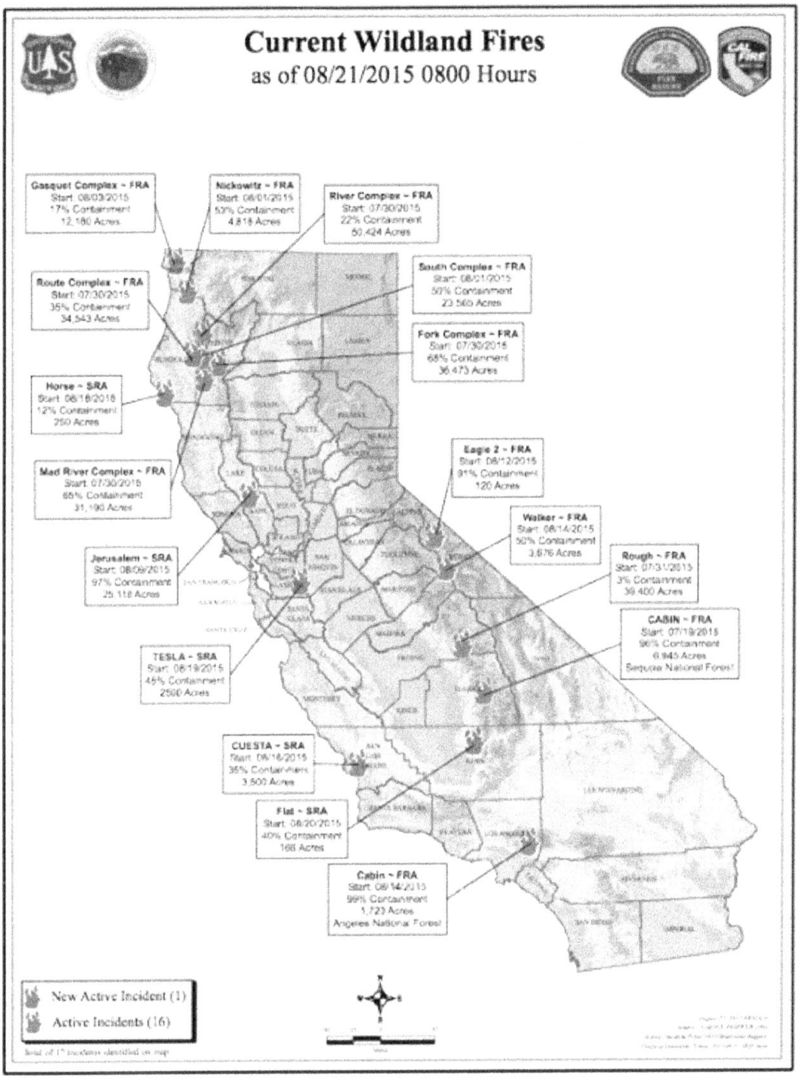


**Map 1 Map of Wild Fires in California 2015**

Today fires rage across the western United States taking human lives and destroying billions of dollars of property as well as the lives of untold thousands of animals. Yet this disaster is human created. My research since the 1960s has shown that since the arrival of humans in the Americas fire has been a tool to shape and tame the natural environment to human uses. With the arrival of Europeans this process accelerated dramatically.

I made a number of core samples to find evidence of carbon from fires in the past and compared the results with published archaeological cores and some geological and botanical studies. It is clear that the evidence of ancient, and especially the record since European colonization, carbon supports the human created fire system we now live in. I published this work in 2002.

This seemed strange given as a student at Humboldt State College in the 1960s I was told by professors in "fire science" that forest fires were natural and had always been happening as long as there were forests. But studies of carbon in the geological record did not support this argument, and a major work in the study of ancient carbon by Edward Golderg (1985) failed to support the assertion. What's more, it is illogical since the only way we could have huge seams of economically useful coal is if ancient forests did not burn and an analysis of coal samples from the oldest deposits bears this out.

## Causes of Forest and Wildland Fire

So if wildfire is not natural then why does it happen? It is true that there are lenses of carbon in the fossil and archaeological record that indicate short term, low intensity fire, but not the massive wild fires that we see today. If this is true, then how do we understand the difference. The exceptions are major asteroid hits and volcanic eruptions. I found that data collected around the world for causes of wild fire show that almost 90% of all wild fire is human caused. This means both people purposely start most fires as well as they do so out of incompetence and stupidity. For the most part humans are fire bugs. We love to see fires. Obviously fire has utility; we cook things, clear land, and it is entertaining and to some people both compelling to watch and beautiful.

Data from a number of sources, including state park services and the federal government cited in my 2002 paper show that where there is human activity we have the greatest concentration of fire and fire frequency as well as intensity. It is true that about 10% of fires are set by lightning and some few by spontaneous combustion, but the grand total can be blamed on humans. However, we hear from the forest industry and most "fire science" researchers that the history of fighting fires has created the fire fuel that produces massive fires. I show in my paper that the history of clearing land by pioneers, of land management by cattlemen and federal and state recreational agencies has routinely fired the forests as well as grasslands to both increase fodder for livestock and to produce more sport hunting and recreational areas. Lumber companies have also regularly set fire to logged over areas to clear debris and these areas are often where new fires appear. The fire regime is human caused and forest fighting efforts over the last 100 years did little to create the situation, usually they were aimed at only protecting property. Nevertheless, we have had a "let it burn" and proscribed burn policy in our wild lands and forests for nearly 20 years and yet the wild fires are getting more numerous and more intense (see chart Climate Central, 2012). I document this problem in a more recent article (Caldararo, 2012).

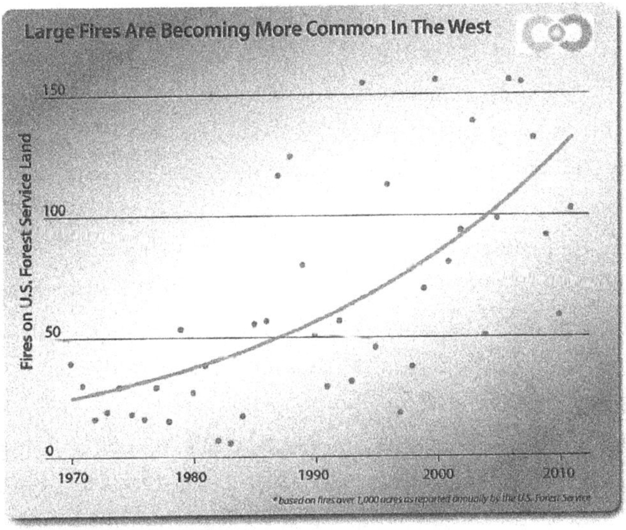


*The annual number of wildfires greater than 1,000 acres on U.S. Forest Service Land has been increasing across the West.*

**Figure 5: Distribution of Wildland fires over time on US Forest Service Land**

Are wild fires now due to global warming? Well we have had global warming episodes in the past before humans existed, both during the early Cenozoic and before, yet no frequency of wild land fires like today. Some argue that climate change (Climate Central, 2012) alone can explain the increase in fires in the past 50 years, but this seems unlikely given the tremendous impact humans have had on the land.

## Solutions to the Problem

How can we stop wild fires? Obviously this is a complex problem, we have a number of industries that have a more than 300 year history of setting fires as part of their routine. This includes **farming and the stock industry**. Farmers and livestock operations can do without the use of fire. This has been shown in California were typical end of the harvest burning of rice fields had largely ended as new products were identified in the waste plant stocks, including pet litter and this has extended to wheat debris. But in many parts of the country this practice continues and has both detrimental health effects to people as well as contributing to general air pollution (for example: http://www.kait8.com/story/13281483/why-farmers-are-burning-fields). Laws to outlaw this practice need to be put in place on a state-by-state basis.

**Fires on rangeland** must stop as well. The use of fire prevents the growth of trees and shrubs and promotes new growth of plants useful for grazing animals. However, it simply creates a fire-adapted ecology promoting a landscape that is formed by fire. However, much of the rangeland that is fired is public lands that are leased by farmers for very little public benefit. The degradation of the land is only a process of erosion and loss of habitat that should be stopped. Dr. Thomas L. Fleischner has documented the political and economic problems of the current situation with unsupervised overgrazing of public lands and the lack of responsibility of the livestock owners to compensate the public for damage (see: http://www.publiclandsranching.org/htmlres/wr_history_politics.htm).

**A stock pond is just one of many alterations ranchers make to the landscape to better accommodate their cattle, to the detriment of native wildlife.**

**Figure 6: Example of degradation of land by over grazing and farming practices, from Thomas L. Fleischner, 2002**

**Logging related fires** to clear debris and to reduce certain types of plants noxious to forest growth should be ended as well as **pesticides** in forests that destroy insects and small vertebrates that feed on undergrowth. But logging in general creates conditions for intense and widespread fires as it promotes "fields" of tree "crops" all planted at the same time so that they grow as a canopy creating crown fires that rage for miles. Logging adapts forests for fire and post-logging fire catastrophes were common in the late 19$^{th}$ and early 20$^{th}$ centuries (see for example: http://www.algermichigan.com/docs/history/MAJOR%20POST-LOGGING%20

FIRES%20IN%20MICHIGAN%20in%20the%201900%27s.pdf). Logging is largely a welfare industry, it relies on subsidies (e.g. public payments for logging roads) for its survival and tax benefits. Some of this has been the result of financial investors as in hedge funds buying up logging lands from defunct companies as in the case of Maxxam. This is described in detail by Richard Widick in his 2009 book.

**Hunting** has a negative effect on reduction of fire burden. As hunters kill browsers the biomass increases year by year. It is no surprise to people that the prairies supported some 30,000,000 to 50,000,000 bison in pre-Columbian times (Isenberg,2000), but that moose, elk and deer existed in huge herds across America. In forested areas there was the wood bison, but the central point here is that the reduction of herds leaves masses of plant life that can become fuel for fire. Eliminating or reducing hunting in parks, and open areas of the plains, prairies and our forests will go far to restoring a balance in growth of plant life and low fire fuel.

**Building in rural areas** new housing is a major problem. Cutting new roads in wild areas for new developments divides native species of animals from their ranges and causes toxic runoff and heat "stripes" across vast areas. Developments set housing into wild adapted ecologies where overextended fire services then are forced to protect homes. Amenities like gardens and trees, usually non-native are introduced. What is usually unknown to most people is that many grasses and brush (like Scotch Broom or French Broom) in our grasslands are non-native and yet are fire-adapted due to their origin in Eurasian contexts due to the long history of slash and burn agriculture (Randall, 2021). Also, many of these imports are not eaten by American wildlife adding therefore to the fire fuel. However, our grasslands (including chaparral) are so fire adapted that burning an area does not reduce frequency of burn, but we see frequent "reburned" areas instead (Sweeney, 1956).

Yet, just the presence of more people who routinely use a grassland area will result in more human-set fire, either intentional or accidental, thousands of structures in the USA are lost every year due to the increasing movement of housing into wild lands and forests (Stein, et al. 2013). This should be stopped or severely limited as the costs of protecting these homes and related commercial developments are unsustainable.

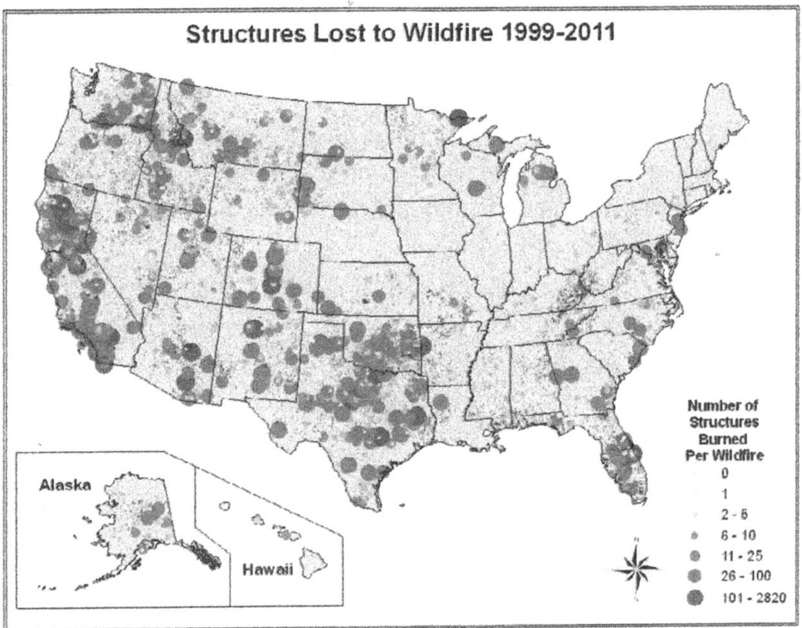


Compiled and mapped by the Fire Modeling Institute, Fire, Fuel, and Smoke Program; Rocky Mountain Research Station; Missoula, MT; 4/5/2012

**Map 2: Structures lost to wildfire from Stein, et al., 2013**

Probably the biggest problem concerning reducing wild fire and its costs involves **recreational use of public lands**. Since we know that people are associated with the greatest number of fires, the presence of people is a factor that cannot be ignored. Yet people want to use the "wild" as a place to enjoy themselves, but they do not want real "wild" animals occupying these areas like wolves, bears, mountain lions or coyotes as someone might get hurt. The fact that our rural areas can hardly be called "wild" today is irrelevant to most. The fantasy of being in the "great outdoors" is a powerful incentive and the recreation industry uses that power to greatest advantage. It would reduce damage to wild lands in both deforestation, survival of wildlife, reduced traffic, pollution and pesticide use, if more recreation took place in a virtual environment as in Second Life or in specially built industrial centers as in Figure 7 for skiing.

Everyone from the Sierra Club to CSAA are boosters for getting out in nature, yet the idea of hordes of people stamping over the last wild space and crushing its natural inhabitants in good healthy exercise is lost to them. The general idea in environmental groups like the Sierra Club is that if people go into nature they will want to protect it. The fact that the more people who do that and the more facilities we build to allow access and overnight accommodations, etc. the less nature there is seems again, just a complication. Mathieson & Wall (1982) discuss the nature of "trampled" tourist sites and the varieties of vegetation that suffer the most. Belsoy, et al., (2012) present a more recent picture of the mass effects of both recreation and tourism on natural areas.

Perhaps the development of indoor recreation is the answer, and support of this trend could be a means of taking pressure off rural and "wild" or once natural areas. Already there are very significant investments being made in experiences like indoor skiing. Another growing recreational opportunity that many young people seem to be attracted to is the virtual experience. Whatever replacement we can create to divert human impact on the open spaces in the future, either to prevent the further destruction of the environment or to limit or outlaw more built space, will have a positive effect in reducing wild fire. The caveat to this rosy picture is that humans have adapted to making and experiencing fire and fire that consumes organic materials on a significant scale. How to repress this trait will be a considerable undertaking. Some mass experience like Burning Man might be devised, only perhaps not as destructive or have as significant effect on the environment as Burning Man (Figure 8), though it might be produced indoors and have a similar effect, given that the first Burning Man took place in San Francisco in a park. Of course, it could also be reproduced in virtual conditions and the human need for fire might also be satisfied in virtual spaces as well.

**Figure 7: Indoor skiing TheSnowCenter.com London**

## Conclusion

From this review of the fire history of the past decade it seems clear that current policy is inadequate to deal with increasing fires, fire intensity, costs of fighting fires and loss of property. The central problem in the past decade is the expansion of homes, roads and recreation into wild areas without planning or cost analysis of how to pay for fires or avoid them. Hunting and grazing should be limited and money should be allocated to educate people to reduce fire potential and loss of life and property.

**Figure 8: Burning man event Nevada Desert**

## References For Part II

Belsoy, Josphat, Korir, Jacqueline & Yego, Jacob, "Environmental Impacts of Tourism in Protected Areas." *Journal of Environment and Earth Science,* Vol 2, No.10, 2012:64-73.

Caldararo N., Human ecological intervention and the role of forest fires in human ecology. *Sci Total Environ;* 2002 Jun 26;292(3):141-65.

Caldararo, Niccolo "Dominant Paradigm of Fire Control: Solution or Problem?" *Forest Research,* v. 1, n. 1, 2012:1-5.

Climate Central, *Age of Western Wild Fires*, 2012: http://culter.colorado.edu /~tims/Wildfires2012.pdf.

Conedera, Marco, Willy Tinner, Christophe Neff, Manfred Meurer, Patrik Krebs (2009) "reconstructing past fire regimes: methods, applications, and relevance to fire management and conservation," *Quaternary Science reviews,* v. 28, 435-456.

Fleischner, Thomas L., "Land Held Hostage: A History of Livestock and Politics," Pages 33-38 in G. Wuerthner and M. Matteson, eds. *Welfare Ranching: The Subsidized Destruction of the American West.* Island Press, Washington, D.C. [http://www.publiclandsranching.org/htmlres/wr_history_politics.htm].

Goldberg, Edward D., *Black Carbon in the Environment*, New York, John Wiley & Sons, 1985.

Isenberg, Andrew C. *The Destruction of the American Bison*, Cambridge, Cambridge University Press, 2000.

Mathieson, A., and Wall, G. (1982). *Tourism: Economic, Social and Physical Impacts.* London: Longman.

Randall, Brianna (2021) "Invasive grasses are taking over the American West's sea of sagebrush," *Science News,* December 6.

Randall, Christina, "Two Years after Waldo Canyon Fire, Colorado Springs Wildfire Mitigation Program in Full Force," FEMA Report: http://www.fema.gov/blog/2014-06-27/two-years-after-waldo-canyon-fire-colorado-springs-wildfire-mitigation-program-full.

Stein, S.M.; Menakis, J.; Carr, M.A.; Comas, S.J.; Stewart, S.I.; Cleveland, H.; Bramwell, L.; Radeloff, V.C. 2013. *Wildfire, wildlands, and people: understanding and preparing for wildfire in the wildland-urban interface—a Forests on the Edge report***.** Gen.Tech. Rep. RMRS-GTR-299. Fort Collins, CO.

U.S. Department of Agriculture, Forest Service, Rocky Mountain Research Station. 36 p.

Sweeney, James R., *Responses of Vegetation to Fire, A Study of the Herbaceous Vegetation Following Chaparral Fires,* Berkeley, University of California Press, 1956.

Syphard, Alexandra and Jon E. Keeley, (2019) "Factors associated with structure 2013-2018 California wildfires," *Fire,* volume 2: doi:10.3390/fire2030049.

Syphard, Alexandra, Heather Rustigian-Romsos, Jon E. Keeley (2021) "Multiple-Scale Relationships between Vegetation, the Wildland–Urban Interface, and Structure Loss to Wildfire in California," *Fire,* v. 4 (1) 12; https://doi.org/10.3390/fire4010012.

Widick, Richard, *Trouble in the Forest California's Redwood Timber Wars*, University of Minnesota Press, 2009.

# Chapter Nine

## Garbage, Waste, Recycling and the "Toilet Mentality"

An article in *Global Change Biology* in 2018 by scientists at the University of Queensland found that most sea turtles have eaten plastic (Duncan, et al., 2018). The contamination of our food is spreading. The problem is garbage and the residues of our behavior, both of which are encased in irrational and illogical thinking. Over 50 years ago George W. Irving, Jr. (1970) published an article demonstrating how the use of pesticides to control for both food production and to established comforts for living (kill mosquitos, termites, spiders, fleas, etc.) conflicted with efforts to protect the environment. Pest control made saving the environment impossible. In much the same way efforts to control wildfire and achieve sustainable logging and paper production as well as maintain or expand nature recreation made programs to preserve wild life and natural environments near impossible. This chapter is about the irrational and illogical nature of garbage, water, seas, air, and how we live in a world built of nonsense. But then we know now that nanoplastic has entered our brains and probably we can use the excuse for our thinking as, "garbage in, garbage thinking."

Yale's Center for Industrial Ecology reports in an issue of *Nature Climate Change* that estimates of garbage going to landfills was underestimated, twice as much is being dumped. One might say that to save ourselves humans should make and use fewer things. The fact that we are toolusers and toolmakers is destroying the earth and burying it in trash.

The report from the Pacific trash "patch" is even worse than people feared given the voyage to the "patch" by Boyan Slat (Jaxon Van Derbeken, "Pacific Trash Patch worse than believed entrepreneur says"

S.F. Chronicle, Aug. 24th 2015) With all the boasting about recycling perhaps it is time to make some hard looks at garbage and make some hard decisions about the future.

As Cverenkarova, et al., show in their 2021 study, plastic waste is now entering our food chain because more plastic is being deposited in the oceans and waterways. This process is clarified by work published by Saeedi (2023), A new study by environmental engineer Jenna Jambeck of the University of Georgia published in *Science*, found that recycling plastic is failing and more than 8.8 million metric tons of plastic are dumped into the oceans each year.

Fraud in recycling is rampant (see reported in the L.A. Times on October 8, 2012 (http://latimesblogs.latimes.com/lanow/2012/10/recycling-fraud-drains-californias-cash.html). Recent studies by MIT's SENSEable City Lab found that the energy involved in transporting recycled materials has not been included in estimates of cost factors in treating garbage. Cost factors and carbon emissions due to transport outweighed the expected benefit of recycling. Amsterdam converts 99% of its domestic and commercial waste into energy that powers its city trams, underground trains and streetlights as well as 75% of the city's household energy needs. A *New York Times* article on Sunday January 11th 2015 reviewed the new garbage burning plants built in the past decade and new ones showing that new technology has made these plants comparable with gas fired energy plants (Timothy Williams, "Plants backed by EPA, but environmentalists point to emissions.")

A detailed study of the actual problems and costs in garbage are found in Edward Humes book, *Garbology: Our Dirty Love Affair with Trash,* presents evidence from several university studies in America and studies by governmental agencies in the European Union that remove the illusion concerning recycling. Reading through *Garbology* you find that greenhouse gases from landfills are now a greater source of pollution than burning garbage. However, new technology has made garbage burning plants clean energy. Changes in garbage and waste to fuel have made garbage a valuable commodity in Europe. The problem with garbage, recycling and landfill sites is growing into a national scandal. In recent years studies have revealed fraud surrounding all three, New York is an example (http://concernedcitizens.homestead.com/FKfacts.html)People can consult the website of the Concerned Citizens of Cattaraugus County, Inc. for details.

The Basel Group in Europe argues that garbage is dumped at sea. Reports by the United

Nations, Group of Experts on the Scientific Aspects of Marine Pollution, have found since their data gathering began in 1990 that more garbage is found in the sea than ever before. Elisabeth Royte's book (*Garbage Land,* 2005) documents this. A report on the world's oceans' health by Markus Eriksen, in *PLOS ONE* and reported in the *Marin IJ* on December 11th 2014, found 270,000 tons of plastic in the oceans and this is degrading into as much as 5 trillion pieces being ingested by fish and aquatic mammals each and every day and then by humans who eat fish.

Landfills can be mined for future fuel. Archaeologist W.L. Rathje found that most garbage does not degrade, especially paper (see his book, *Rubbish!: The Archaeology of Garbage,* 2001).

The most comprehensive article on burning garbage is by Bruce Piasecki, David Rainey and Kevin Fletcher, published in 1998. Innovations since 1998 have made these plants even cleaner (Chaowiang, et al., 2022).

Building co-generation plants can be amortized by fees collected for garbage disposal residents are already paying. Due to cheap oil renewable energy is no longer economically feasible.

Burning garbage can end expansion of existing dumps. Dumps are expanding all over the country and abroad, impacting ground water by increasing pollution in runoff, producing more greenhouse gases: methane and CO2 production. New York transports over 10,000 tons of garbage to landfills each day, other major cities follow closely. Europe is building more garbage to energy plants each year (Elisabeth Rosental, *New York Times*, April 12, 2010). In the USA in 2001 only 14% of solid waste is burned, 33 million tons out of 229 million (http://www.epa.gov/cleanenergy/energy-and-you/affect/municipal-sw.html). The EPA estimates that about 30% of all solid waste is recycled in the USA, thus burning and recycling leave over 50% to be placed in landfills or exported abroad.

## References

Cverenkarova, Klara, Martina Valachovicova, Tornas Mackulak, Lukas Zernicka, and Lucia Birsova, "Microplastics in the food chain," Life (Basel) Dec. V. 11, n. 12, 1349: doi: 10.3390/life11121349.

Duncan, Emily M., Annette C. Broderick, Wayne J. Fuller, Tamara S Galloway, Matthew H. Godfrey, et al., (2018) "Microplastic ingestion ubiquitous in marine turtles," Global Change Biology, v. 25, n. 2, https://doi.org/10.1111/gcb.14519.

Chaowiang, Suntaree, Vishakha Sing, Pornthep Chiraprawattrakun and Sumet Tangprasert, (2022) "Key points and best practices for successful municipal solid wast incineration," *The 6th International Conference on Energy and Environmental Science,* IOP Publishing, doi:10.1088/1755-1315/1008/1/012020.

Irving, Jr. George W. (1970) "Agricultural pest control and the environment," *Science,* v. 168, 1419-1424.

Jambeck, Jenna R., Roland Geyer, Chris Wilcox, Theordore R. Siegler, Miriam Perryman, et al., (2015) "Plasti waste inputs from land into the ocean," *Science,* Feb 13, v. 347, n.6223, 768-771.

Piasecki, Bruce, David Rainey and Kevin Fletcher, (1998) "Is the combustion of plastics desirable?," American Scientist, v. 86, n. 4, DOI: 10.1511/1998.31. 364.

Saeedi, Meysam, (2023) "How microplastics interact with food chain: a short overview of fate and impacts," *Journal of Food Science and Technology,* https://doi.org/10.1007/s13197-023-05720-4.

# Chapter Ten

# Conclusion and Perhaps a Positive Vision of the Future

Most visions of the future are dystopian. The ruins are given as the result of human failure to control greed, or to extraterrestrial intervention either by asteroid or aliens. The one provided here begins in that manner as it is a simile of factors now functioning to bring about disaster, both climate change and thermonuclear war and perhaps even human created pandemics. The purpose of this exercise in creative history is to outline in more brilliant colors the frame of the future. This is a future that, at its end, to me is positive and sustainable. Even futurists like Khanna (2016) count on massive human migrations and modifications of social organization to adapt to severe changes in weather.

We need more positive visions of the future. It seems they are more difficult to conceive and write than those fashioned in terror. Some are quite interesting like the *Golden Compass*. The imagining of future ways humans might live is necessary to provide people with alternative ways of constructing their realities and their identities. I call the one that follows, Past-Future-Perfect. It considers the use of past ways people have lived successfully, modified by present technology and desires, then projected into dimensions of our hopeful realm of lifeways.

## Reference

Khanna, Parag (2023) *Connectography: Mapping the Future of Global Climate*, New York, Random House.

# THERE ARE NO KINGS IN BASQUE COUNTRY

By

Nelson Ogalala Chan

a fragment from his autobiography:
"Reverse 1492: The Re-Indigenation of the World"
Edited by Niccolo Leo Caldararo

# A note from the editor...

I met Nelson Chan In 1975 in a village south of Cuernavaca. My girlfriend and I had been ejected from a bus because the ticket we had was not properly stamped according to the bus driver. Nelson was sitting in the plaza near where the bus stopped to drop us off and heard our complaints. As the bus left without us, he waved us over and tried to comfort us with an explanation of what had happened: The bus driver operated as an independent agent and would not be paid for tickets the ticket agent had sold but cleverly failed to stamp, given that he felt two tourists would not come back and file a complaint so he could pocket the money.

As we waited for the next bus, Nelson told us his life story, or stories, since he had so many adventures. His physical appearance seemed to place him in his thirties, but he had a certain world-wise nature that made him appear much older, his ready wit and vivacity gave him a younger mien. He had been a sailor, a student in chemistry, a travel agent and writer for travel magazines, railway worker, all jobs in countries across the world in numerous countries. He asked us what we did, and when I told him I was an anthropology student, he began to speak of different cultures he had come into contact with, and then, when my girlfriend told him she was trying to be a writer/artist, he spread before us his tales of travel and then some of his experiences with peyote and time travel. Last of all, before the next bus to Mexico City arrived, he began to relate his history of the world as a means of healing the planet. Nelson argued that we had to create a history that would provide for future generations to survive. His history was one he saw born in the most horrifying events and I outline these in as much of his "future history" as I can recall below. We did take our tickets back to the station and the clerk duly stamped them. "Sorry," he said as he smiled at Chan...

The reason I am publishing the story now is because someone sent me an outline of it two years ago. I had heard from Nelson earlier, in 1978 I think when he sent me a story he had written about changing into a Jaguar. It was from the Jaguar's point of view, like the Roman book, *The Golden Ass*. I do not recall giving him my address when we met, but we did tell him about the Goodman Building where we were living, and since it was somewhat famous among artists at the time (it was a live/work squat where artists, historians, actors, musicians, etc. from all over the world came to work) it is not hard to imagine he came across its location. I lost that story, and do not remember much about it. However, if anyone knows Nelson, please show him this story. If he does not like how I remember it, have him contact me.

I say, "someone" sent me the outline as the envelope it came in was without return address or name. It might have been the girl I was traveling with when I met Nelson. She passed away about the time the envelope arrived.

Niccolo Caldararo

# There are no Kings in Basque Country

1

Once when I was a child my grandfather, took me to Gernika to see Gernikako Arbola, the tree of the Basque people. It was at the time when chaos had engulfed the world, before the Orchard had come. He had said that in each person's life things seemed normal, even war, as all things become normal in time and people come to adjust to how they are. The Tree, he had said, represented the people, the branches and the leaves, the families, the trunk, the history and the roots were the past of generations unknown, when to be a Basque came into being. The Basque must be thankful for those in that past who made being Basque.

From time to time I came across books about the Basque, like Mark Kurlansky's the *History of the World According to the Basque,* where the Basque fierce independence and democracy are central to their story. Other books, like Roger Collins' *The Basque*, describe the mists that cover the obscure origins of this people. Then there was the Movie, To Die In Madrid, which depicts the fight of the Basques against fascism. It seemed horrific when the Nazis purposely bombed the Gernikako Arbola and the town of Gernika. But from an acorn a new Gernikalko arbola grew and the Nazis were no more. I kept these ideas as a foundation to understand my own origins and path in the world.

Things changed when my grandfather was a child, from a time when there were nations and those nations made of people who traded and fought as groups. My great grandfather was Chinese, born in Buenos Aires, but raised in Chicago. His mother was a Native American woman in New Orleans. He came to France with the American army in WWII and stayed in Bayonne as a civilian. He married a woman from Carcassone whose mother was a Tuareg from Mali and her father a man of Langue d'oc. My grandfather had been educated in Macao but worked in Japan, his wife was a photographer who fell ill in Paraguay and died in Rio. My grandfather's work took him to Taiwan where he married a woman of Okinawa. They had returned to Bayonne to care for my great grandfather. My father lived in the Philippines and married a woman from Luzon. My wife, a woman from southern India, a vet who specialized in small animals.

But when my grandfather became a man, the world became the prisoner of sects and fanatics who joined to destroy borders of nations and differences in beliefs. They set fire to all things that displeased them and murdered those with whom they disagreed. Yet this caused war and destruction so that weapons of all kinds were used, including nuclear ones and violent swarms of fanatics covered the earth until there were no farms or factories. The earth became a wasteland.

Globalization – the civilization of the modern era - had brought us to this place, where the contributions of millions of minds had opened great vistas of possibilities, but with it came a terror of certainty so unrelenting that the light of the world faded and the way of being human was lost.

In this wasteland of violence rose one group who enslaved all others and swept the continents of disorder until there was only one nation, the Orchard. The Basque did not succumb to this terror or the fundamentalism that drove it, but remained within their countries. My grandfather had come there with his sister and wife, and he told of travelling across the Orchard's roads where people did not live as in cities, but where the buildings had been destroyed and the roads and highways were now places filled with shacks, as in Medieval Europe after the fall of the Roman Empire. The Basque kept their roads clear, their cities tidy and they kept farming and making things. It was, my grandfather had said, as if most people on the earth had listened to a song and become mad, but the Basque could not hear the song and so kept being Basque. They met the crisis by meeting and discussing the evidence, each person a voice, no leader, no chief, only a stream of individuals examining the cosmos together. It was as if the mists of the past still surrounded the Pyrenees, mists so old they still recalled the days before the coming of man, before there were Basques. Then there were no Basque, only people, then they became Basque.

There were refugees, always at all times, even before the Romans came. These were people who came to Basque country, searching for aid and peace, to learn Basque and be. But when the Orchard was victorious, the flood of wounded and hungry became a desperate mass. Unlike in the past where some could stay and be productive and Basque country grew. Now there was no where to go. Some tried to go into what had once been the lands of Gascogne after the fall of the Roman Empire, and to Catalonia and the Cantabrias, for there were no people there any longer, and Barcelona was a ruin.

The Basque never had kings, were perhaps matriarchal and are said to never have conquered others. They tended their garden, though were often forced to fight for others to survive, in the Roman Armies, and in the service of the Visigoths and Franks. There were no mountains among them, no leaders who dominated the landscape of a nation, not in the pass where Roland fell or in the blood soaked fields of the Ring of Iron before the fall of Bilbao in the Civil War. "It goes," said the woman who sold flowers in the market in San Sebastian, "from large to small and back again, it grows but stays the same. 'That's not progress, and does not make history for a nation,'" they say. She shrugged. There is a place in your heart that is carried across the generations. Whether it be in Spain or America, Russia or Australia. You come to a clearing in the forest and you know you are home. That is "Basque country." It is your country, because you are there.

*eliefs in mana—a supernatural force or power, which peopl*
*ay manipulate for their own ends—are widespread. Mana*
*in reside in people, animals, plants, and objects, such as*
*ie skull held here by a member of the head-hunting Iban*
*ibe of Malaysia.*

2

To the rulers of the Orchard the existence of Basque country was, at first, an opportunity. They bought food and manufactured goods. But the Basque would not sell them weapons, vehicles, ships or airplanes. Those that existed outside Basque country were breaking down as there was no education or training in the chaos of the Orchard. Teaching and learning were forbidden to them, there was only subjection and taking, knowledge was evil to them. Eventually the Orchard sent its armies against the Basque and the Basque defeated them, but then more came by the 100s of thousands, month after month until the Basque were driven to their mountains. Then the storm seemed to lessen, week by week the numbers of attackers fell until what had been a torrent became a stream and then only stragglers came asking for asylum. They told of great armies of the sky who destroyed the Orchard cities and armies.

As the days passed unease set in after the initial relief as there was no understanding of what force had set upon the Orchard and if it would be a danger to the Basque. Planes were seen flying over the countryside that had once been battlefields. The Orchard had no planes so the new force was a mystery. One sunny day radar picked up a number of planes heading for Bilbao. Broadcast to the Basque media came a message from the Japanese, that an emissary was on the way. The Basque air force moved to intercept, but there were no aggressive acts. A single plane landed in Bilbao.

The Mayor of the town and a mass of Basque came to meet this person. Out of the plane came a woman in a kimono followed by a number of men in traditional Japanese dress. The woman came to the Mayor and said in Basque:

"We suffered the first horror of nuclear war decades ago. Modernity produced weapons of mass destruction and death-worshiping dogmas. All the ideologies of the world are dead. All religions false. We endeavor to destroy man and bring peace. Do not expand, be Basque." She turned and they returned to the plane and left. It seemed obvious the Japanese were also immune to the music of the sects.

'Woodpecker finch, *Cactospiza pallida*, using a ool [here a twig] to extract insect larvae from a lead branch of *Bursera graveolens*." [From *Ecology* *ınd Evolution of Darwin's Finches*; photograph by R. Perry]

3

When I was a child we often heard that the Japanese had pursued the fanatics across the globe killing all humans they encountered. They pursued the fully globalized, those lost to technology who like zombies walked the streets staring at glowing screens, locked into the message of the Orchard still broadcast from robot servers hidden in the depths of ruins. Purifying the land as they went, one might say, following the signals of the living dead. Where no "modern people" were only traditional people, as in jungles of the Amazon or the Congo, they did not go, but where the plague had passed of the Orchard they spread death. But among the Lapp, the Hadza, in parts of Indonesia, there remained some clothed in a protective shield of ancient culture. Still here and there, in America, in Australia, and other places were people who had resisted the song and were living without the power that had been the Orchard.

A quiet grew as time passed.

Basque fishermen and women had for thousands of years before the Romans arrived, fished across the Atlantic without contaminating other peoples. Now they brought reports of huge schools of fish, of great groups of whales and of forests in the Americas and Europe. The world was healing. This meant little to the Basque who just lived their lives as they had for thousands of years before. No more people came asking for sanctuary. The Japanese did not come again. The land was quiet except for the skies being filled with birds of all kinds.

Radio messages from Ainu Ham Radio operators in Japan told of terrible riots by fanatics and it appeared that the Japanese were also not immune to the disease. Decades went by with no more broadcasts until a Basque ship landed in Hokkaido. The country had returned to nature and the Ainu had returned to the old ways, like the Andamanese, they desired no contact and turned their backs on the visitors.

Some people continued to go into the lands outside Basque country to hunt or fish. Some went to the ocean outside Catalonia, but none stayed. My wife and I lived in the village of Ainhoa with a house full of cats on the top of the mountains. Buried beneath the house are the bones of people who lived before us, some known, some unknown, for all people are our ancestors. We read their lives and learn how to live so there will always be enough and peace can reign not men. As the Japanese had requested, the Basque did not expand, but then the Japanese were no more and yet it did not matter. The world was at peace and the Basque remained Basque.

"But father," asked one of Nelson's children during his telling of this tale, "Was it instinct or genetics that saved the Basque?"

"No," Nelson responded, "It was the culture of the indigenous people that was like a protective bubble against the madness of the Orchard."

"We, and all people who survived, are a coat of many colors. That is the only way the peoples like the Basque and Ainu could have taken in so many refugees and resisted the madness together."

## References

Pullman, Philip (1995) *Northern Lights*, London, Scholastic UK. (Made into a movie titled, *The Golden Compass.*

# Appendix

List of initiatives people & governments could take to produce a more sustainable Earth.

1. Stop having babies, if the fertility could be reduced to 1 child per 10 women that would make a health world more likely. 1 in 100 better. There are too many people now for humans to ever attain sustainability without sacrifice. Women who do not have children should be treated like the heroes they are.
2. Reduce living units in developed rich countries to 500 sq ft per person that would reduce heating and cooling and lighting.
3. Move humanity from hot, humid, and cold areas to the temperate zones that will save energy in cooling and heating.
4. Burn garbage increase recycling today too much plastic and toxic materials go into the ocean, seas and lakes.

www.ingramcontent.com/pod-product-compliance
Lightning Source LLC
LaVergne TN
LVHW010718210926
848485LV00008B/1450/J
*9781680534160*